Wood Flooring

A COMPLETE GUIDE TO

Layout, Installation & Finishing

Wood Flooring

CHARLES PETERSON

WITH ANDY ENGEL

The Taunton Press

The Taunton Press, Inc., 63 South Main Street,
PO Box 5506, Newtown, CT 06470-5506
e-mail: tp@taunton.com

EDITOR: Peter Chapman
COPY EDITOR: W. Anne Jones
INDEXER: Jay Kreider
COVER DESIGN: Nick Caruso
INTERIOR DESIGN: Nick Caruso
LAYOUT: Nick Caruso and Amy Griffin
ILLUSTRATOR: Vincent Babak
PHOTOGRAPHER: Randy O'Rourke, except photos on front cover (bottom): © Tom Hopkins Studio; pp. 8, 9, 13 (bottom), 20, 21, 23 (top, top middle, bottom), 24 (bottom left, right), 28 (bottom left), 30–32, 45, 46–47 (left), 48, 56, 69–71, 91 (top left and bottom left), 92 (bottom), 93, 96, 98–100, 104 (bottom), 105, 108, 110–111, 113, 114 (photos 1, 2, 4), 115 (top right, middle right, bottom left, right), 118–119, 120 (left and right), 123 (middle, bottom), 125, 130 (top left, bottom left, bottom right), 134 (bottom left), 137, 138, 151, 153 (top), 158, 159 (top photos), 162–163, 165 (bottom), 168, 172, 178 (top), 181 (top photos), 182–184, 186, 200, 201 (top), 219 (bottom), 231 (left), 233–235, 244 (right), 247, 259, 262, 265, 266 (bottom), 267 (top photos), 270 (top), 274 (left), 276–277, 279, 281–284, 286, 289, 290, 292–294, 295 (top), 296 (left), 297, 298–299, 300–304 (top, bottom), 305–307, 310–311: Courtesy Charles Peterson; p. 22: Courtesy Owens, Inc. (www.teamowensinc.com); p. 23 (bottom middle): Courtesy Andy Engel; p. 27: Courtesy Heart Pine Company (www.heartpine.com); pp. 41, 141, 165, 180 (bottom), 249: Courtesy *Fine Homebuilding* magazine © The Taunton Press, Inc.; pp. 109, 314 (top left): © Thomas Jefferson Foundation; pp. 114 (3), 123 (top), 133, 179, 244 (left), 313, 314 (top right and bottom left), p. 315: Courtesy Universal Floors (www.universalfloors.com); p. 131, 266 (top): Courtesy Decorative Imaging (www.decorativeimaging.com); pp. 295 (bottom), 296 (middle and right), 304 (middle), 308 (top): Courtesy Timothy K. Moore; pp. 316–323: Courtesy Andy Rae except p. 316 (bottom middle and bottom left), 318 (middle), 319 (middle), 320 (right), 322 (bottom and right): Courtesy Charles Peterson

Library of Congress Cataloging-in-Publication Data

Peterson, Charles, 1962-

Wood flooring : a complete guide to layout, installation & finishing / Charles Peterson.

p. cm.

Includes index.

ISBN 978-1-56158-985-2

1. Flooring, Wooden. 2. Parquet floors. I. Title.

TH2529.W6P48 2010

694'.2--dc22

2009047189

Printed in the United States of America
10 9 8 7 6 5 4 3 2 1

Working wood is inherently dangerous. Using hand or power tools improperly or ignoring safety practices can lead to permanent injury or even death. Don't try to perform operations you learn about here (or elsewhere) unless you're certain they are safe for you. If something about an operation doesn't feel right, don't do it. Look for another way. We want you to enjoy the craft, so please keep safety foremost in your mind whenever you're working wood.

To the National Wood Flooring Association, and their commitment to education

ACKNOWLEDGMENTS First, I want to thank my wife, Elizabeth, for her continued love and support. For five years, she has put up with photo shoots and partially built floors, test panels, and a mountain of books and papers in the middle of her home. Thank you to my sons, Chad and Clark, who were instrumental in helping simplify and test my ornate flooring techniques. They helped build many of the inlays featured in this book. Thank you to my mom and dad, who worked years to make their son an engineer and then had to stand by and watch him play with sawdust. And to my grandfathers: Charles Peterson, who nurtured my passion for writing and teaching; and Webster Gibbs, a World War II hero who told me you have to do what you love in life or it is not living.

More people than I can name helped to make this book. I'd especially like to acknowledge my friend Patrick Van Dyke, who magically organized rooms before photo shoots and offered up a seemingly endless string of great ideas. My humble friend Frank Kroupa graciously shares his wealth of knowledge with craftspeople all over the world. Don Connor on more than one occasion helped me verify information, and Bryan Liebenthal came to the rescue in helping trim out the "Floor of the Year" room that appears throughout this book.

At The Taunton Press, Peter Chapman was an incredible editor and provided me with the best possible team to work with. Randy O'Rourke is one of the finest photographers in the country and someone I had always hoped to work with. His eye for composition and lighting is unparalleled. My coauthor Andy Engel had the daunting task of fitting thousands of pages of my information between the covers of this book. His work relieved the dryness of my technical style of writing and focused my information. Andy's knowledge as a craftsman provided hundreds of questions that added great depth to the book.

My friend Tim Moore, Director of Technical Services of Moore & Neville, Inc., provided many photos of flooring problems in chapter 12. Mike Sullivan dropped everything else in his busy life to give me walls and a ceiling for a photo shoot, and Sprigg Lynn, a true gentleman and a great craftsman, provided many of the photos of historic floors in this book. His company, Universal Floors in Washington, DC, is one of the world's foremost in wood flooring restoration, and I have enjoyed consulting and preserving history with Sprigg. And last, my best guy friend Dana A. Neale, whose opinions I value greatly.

Contents

Wood is such a beautiful medium. It comes in an unlimited palette of colors, it bends, and it can be cut and glued into any form. I worked on my first wood floor in 1978, and I've been hooked ever since.

Until the 1960s, wood was the flooring of choice. At that time, several factors combined to dethrone hardwood. First, nylon-tufted carpet hit the market and the cost of wall-to-wall carpet dropped by nearly half. Suddenly cheaper than hardwood, carpet became a fashion trend. Then, in 1965, carpet was approved for homes with FHA-backed mortgages. The coup de grâce came during the winter of 1972–73 when heavy snow in the north and heavy rain in the south stopped logging. With no new raw material, the price of wood flooring skyrocketed. Hardwood flooring production went from over 1 billion sq. ft. in 1955 to about 99 million sq. ft. in 1975. This huge drop in demand caused most wood flooring master craftsmen to find different work or retire. Along with them went access to their knowledge.

Forty years later, wood is again becoming the floor covering of choice. Unlike carpet, wood flooring can easily last the lifetime of a home, and its durability is coupled with beauty and warmth. It is both renewable

and recyclable. Normal maintenance is sweeping or vacuuming. Stains can easily be cleaned from it, and wood flooring does not collect dirt and contaminants that may have negative health effects. With wood species and products in every price range, more wood flooring options are available than ever before. Today's finishes are more durable and the adhesives are stronger. In addition to traditional solid hard-wood, we now have prefinished engineered wood floors, which combine the stability of plywood with the durability of hardwood.

I fell for wood flooring at a time when the market for it had collapsed. To make a living, I pursued a career in engineering, though wood flooring remained my passion and consumed all my free time. For the past three decades, I have read every bit of information on wood floors that I've been able to find, and it hasn't always been easy. Finally, a decade ago, I was able to retire from engineering and devote myself to wood flooring.

Since then, I have won international competitions for wood flooring craftsmanship. I have combined my

graduate work in engineering and wood science with my passion for the art of wood flooring to become an internationally recognized wood specialist. I donate my time and knowledge teaching and helping to raise the quality of the industry. I contribute to industry technical manuals and many national publications, and I'm involved in an ongoing effort to create national certification programs for craftsmen and inspectors.

I hope this book will aid professionals to elevate their expertise in wood floors and help resurrect the master craftsman of yesteryear. Frank Lloyd Wright once said, "We may use wood with intelligence only if we understand wood." This book provides the basic knowledge needed to understand wood flooring and how it behaves and includes the treasured techniques of the world's leading wood floor craftsmen. These techniques have been simplified so that anyone with basic carpentry skills should be able to create wood floors befitting a palace. Finally, the book covers problems that can occur with wood floors and their solutions. Consider this book your comprehensive resource on wood floors.

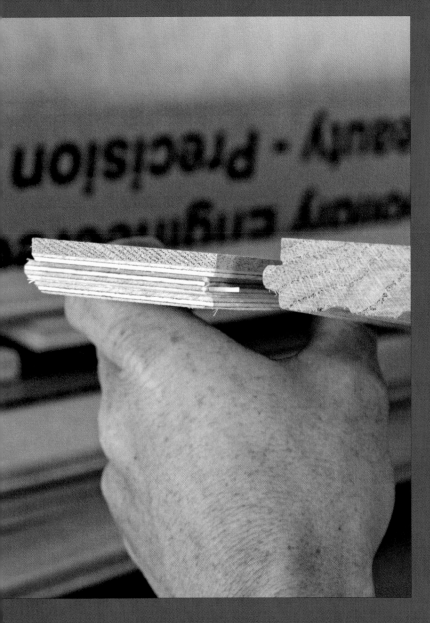

CHAPTER ONE

Wood Basics

INSTALLING A WOOD FLOOR REQUIRES A SUB-stantial investment in time and material. Done correctly, the floor's warmth and beauty will be unmatched and can last a century. Unfortunately, flooring is often installed poorly. While it may look great initially, it can fail prematurely. Not only is the cost of the installation wasted, but fixing problems can also be costly and render the building virtually unusable for the duration of the work. Repairing failed wood floors may require removal of the old floor, installing a new floor, removal and storage of furnishings, temporary housing for the homeowners, and reimbursement for loss of use of the dwelling.

When wood floors fail in an existing house, the expense can go beyond tearout, replacement, and finishing. Furniture often must be moved, while walls, trim, and the owner's possessions must be protected. Sometimes, the owners must be relocated for the duration of the work.

Most wood floor failures are preventable with a basic understanding of wood and how it behaves. Most wood floor problems have to do with moisture. Nonetheless, people often attribute the cause of the problem to "bad" wood. They find it hard to believe that moisture can affect wood that has been kiln-dried. Whether moisture comes from liquid water or vapor in the air, wood flooring will soak it up like a sponge. At a certain point, the wood changes dimension.

Moisture can cause a wood floor to expand to such an extent that it actually moves the walls of a building. It takes over 1,000 lb. per sq. in. to crush the wood cells of a red oak board, yet many oak floors that have failed because of moisture-driven expansion have permanently crushed boards.

Hardwoods vs. Softwoods

Lumber is generally categorized as hardwood or softwood, but don't let the names confuse you: Some softwood flooring can be harder than some hardwood flooring. Seeds and leaves, not their working

Thousands of years ago, Egyptian masons understood the power of wood and moisture. They drove dry wooden wedges into holes drilled in rocks, and then soaked the wedges with water. The force of the swelling wooden wedges split mountains of stone.

When wood flooring expands due to moisture absorption, it can push hard enough to move walls and crush the edges of boards that are capable of withstanding 1,000 lb. per sq. in. of pressure.

characteristics, distinguish hardwood trees from softwood trees. Hardwood seeds have some sort of covering. This might be a fruit, such as an apple, or a hard shell like an acorn. Hardwoods are broad-leaved trees and typically lose their leaves during cold weather. Softwoods are conifers and grow seeds in cones. Conifers' foliage resembles needles, and most conifers keep their needles year-round.

Trees grow outward by making new wood just under the bark in the cambium area. As they grow, trees create growth rings, which indicate varying degrees of cell growth. In temperate regions, trees generally experience vigorous growth early in the season, which produces springwood or early wood. Denser, darker summerwood or late wood is produced in the slower growth period later in the season. Because they coincide with annual seasonal changes, these rings are called annual growth rings. In floors, we see these growth rings as the wood's grain.

Newly formed wood is called sapwood. It contains living cells and helps to move sap through the tree. As the tree grows outward, the innermost sapwood dies and stops conducting sap. When this happens, it becomes heartwood. As sapwood transitions into heartwood, the

PARTS OF A TREE STEM

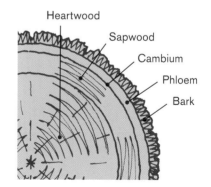

Heartwood

Sapwood

Cambium

Phloem

Bark

Free water fills the spaces between
wood cells and evaporates relatively
easily, whereas bound water chemically joins to the cell walls. This bond
makes it more difficult to remove
from wood. This bound water is what
causes wood to shrink and swell.

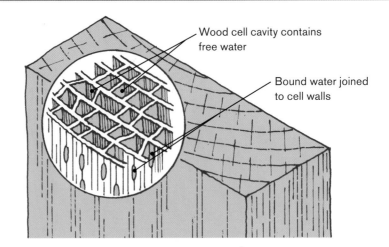

Wood cell cavity contains
free water

Bound water joined
to cell walls

wood can darken due to extractives, which are minerals and natural
chemicals that deposit into the cell cavities or infiltrate into cell walls of
the heartwood.

Extractives contribute characteristics to the wood, such as color,
smell, density, resistance to decay and insects, hardness, specific gravity,
and sensitivity to moisture changes. Some woods such as teak contain
extractives that impede the adhesion of glues and finishes. Because the
extractives plug small cell wall openings in the heartwood that would
otherwise pass water, sapwood is generally more sensitive to moisture
changes than heartwood. Flooring that's all heartwood is usually less
susceptible to moisture-related dimension changes.

Wood Dries in Two Stages

Living trees are full of water. The water in a freshly felled tree generally
outweighs the wood. We describe wood moisture content as the weight
of its water divided by the oven-dry weight of the wood and expressed
as a percentage. The moisture content of wood from a newly cut tree can
actually be above 100%.

Wood holds moisture in two ways. The majority of it is free water,
held in the spaces between the wood cells by capillary action. The
second way that wood holds water is through a chemical bond. So
called "bound water" is attracted to hydrogen-bonding sites in the wood
cells. In simple terms, the oxygen atoms in the cell walls have a slightly

negative electrical charge that attracts the slightly positively charged hydrogen atoms in the water.

Drying wood does not shrink until all the free water is gone and the bound water starts to come out of the wood cell walls. The transition point in the drying process at which the cell walls are completely saturated with bound water but the cell cavities contain no free water is called the *fiber saturation point (FSP)*. Wood cannot change dimension until its moisture content falls below the FSP and water is lost from the cell walls. This is important in how wood dries and how it behaves as flooring as it absorbs moisture from the air or the structure.

The FSP varies between species. Wood species that contain large amounts of extractives generally have a lower FSP. For example, the FSP for southern yellow pine is 29%, while for rosewood it is 15% moisture content. In most wood, the FSP is between 25% and 30% moisture content. Wood used in a conditioned indoor space will eventually dry to well below its FSP, continuing to shrink in size until it reaches a point where it's at equilibrium with its environment. The point to drying wood is to bring it to that equilibrium moisture content prior to installation, keeping further dimensional change to a minimum.

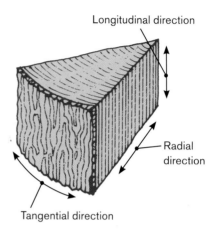

Longitudinal direction

Radial direction

Tangential direction

Wood flooring expands when it absorbs moisture, so it's critical to leave a large enough gap between the flooring and all vertical obstructions such as trim, columns, tile, or cabinets to protect them from damage.

Dimensional Change

When wood is below its FSP, moisture change causes it to expand and contract in varying degrees depending on the orientation of its growth rings to the surface of the board. Wood shrinks most along its growth rings (tangential shrinkage), and least radially to them (radial shrinkage).

Because of the differences between tangential and radial shrinkage, how boards are sawn from the log affects their behavior during moisture changes. The growth rings of a plainsawn board are oriented 0° to 45° to the surface of the board, which is tangential to the growth rings. The farther a plainsawn board is from the center of the log, the more tangential to the growth rings

WOOD MOVEMENT IN PLAINSAWN LUMBER

When changing moisture content causes wood to grow or shrink, the degree of the change depends on the grain orientation. The greatest change occurs tangent to the growth rings.

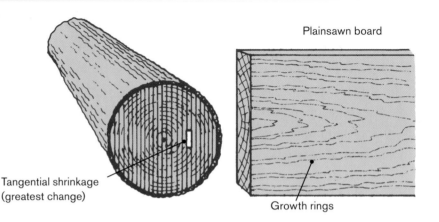

Plainsawn board

Tangential shrinkage (greatest change)

Growth rings

WOOD MOVEMENT IN QUARTERSAWN LUMBER

Quartersawn lumber takes advantage of the fact that wood changes direction least radially to its grain direction.

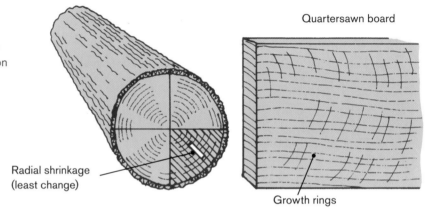

Quartersawn board

Radial shrinkage (least change)

Growth rings

its faces become, and the more they expand and contract in width with changes in moisture. This means that plainsawn flooring will shrink most in width. Plainsawn wood is also commonly called flatsawn or flat-grained.

The growth rings on quartersawn boards are oriented 60° to 90° to the surface of the board or radial to the growth rings. Quartersawn wood experiences radial shrinkage along its width, and so varies less in width with moisture changes than plainsawn wood. A plainsawn red oak board may shrink and expand more than twice as much across its width as a quartersawn board. Quartersawn wood is also commonly called vertical grain or edge grain.

The least moisture-related dimensional change in wood occurs along its length. Longitudinal shrinkage from green to oven-dry is only

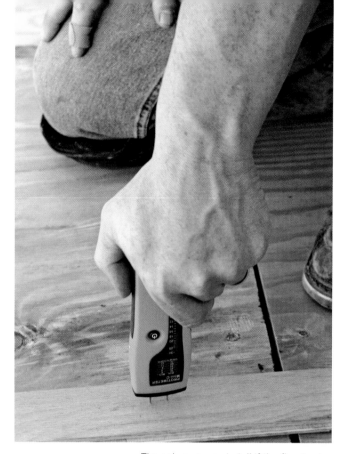

0.1 to 0.2% for most species of wood, while tangential shrinkage might be as high as 8.6%. Longitudinal shrinkage is generally ignored in any prediction of wood behavior. Engineered flooring, which consists of a substrate of plywood topped with a relatively thick veneer that shows as the finished floor, moves about the same amount in all directions.

The amount that wood moves isn't just dependent on how it is sawn, or whether you're measuring along or across the grain. Different species of wood move more or less than others (see Appendix A on pp. 316–323). To allow flooring to expand if it absorbs moisture, it's critical to leave a gap between it and obstructions like walls.

The only sure way to tell if the flooring is at the correct moisture content for the job is to measure it with a moisture meter.

Acclimating Flooring to Its Environment

Because of how wood swells and shrinks with changes in moisture content, most wood flooring problems can be attributed to moisture. The best way to prevent wood flooring problems is to use a moisture meter to make sure that the flooring is at the optimum moisture content for its environment before installation.

The dimensions of wood flooring will change depending on the relative humidity of the surrounding air. The relative humidity is the amount of water vapor in the air at a given temperature, expressed as a percentage. The water contained in both wood and air exerts a vapor pressure. The higher the relative humidity, the higher the air's vapor pressure. Similarly, the wetter the wood, the higher its vapor pressure. If the water vapor pressure in the air is lower than the vapor pressure within wood, wood loses moisture (*desorption*). The reverse is true as well (*absorption*). The bigger the difference between the vapor pressures, the more rapidly changes occur. When the vapor pressure within the wood equals the vapor pressure in the surrounding air, no more desorption or

If wood flooring is too wet when installed—often because it picked up excess moisture from construction sources such as drying concrete, plaster, or paint—gaps will open up as it dries.

Floor finishes add a layer of wear resistance, but, perhaps as important, they also help protect flooring from moisture by slowing down its absorption and drying. This evens out occasional drastic atmospheric relative humidity changes.

Moisture can enter a floor from a damp cellar or crawlspace below. Using a vapor retarder such as Fortifiber Aquabar "B" below wood flooring helps to protect it from such sources of moisture.

absorption occurs. This point is called *equilibrium moisture content*, or EMC.

Homes under construction go through a range of moisture levels. Since wood is sensitive to changes in moisture content, it makes sense that wood flooring should be one of the last items installed. It is advantageous to install the wood flooring with a moisture content that is as close as possible to normal living conditions. Wood flooring that gains moisture on the jobsite prior to installation may develop big gaps between the boards as it loses moisture during the dry season of the year. Wood flooring that picks up moisture after installation may cup or buckle. Flooring that buckles actually detaches from the subfloor. It is common to see flooring lifted 1 ft. into the air. I have seen construction dampness expand wood flooring enough to move walls, doorframes, and cabinets.

Short fluctuations in relative humidity usually have no appreciable effect on wood moisture content. Over time, flooring will stabilize at an equilibrium moisture content dictated by the environment. Floor finishes help moderate this process. Three coats of most finishes help to exclude moisture for a few days, but their effectiveness drops off dramatically over a couple weeks of high humidity or dry air. Moisture can also come from beneath the wood floor. A vapor retarder such as

EXPECTED INTERIOR MOISTURE CONTENT

Wood's moisture content varies with location and the time of year. The map at right shows the expected annual swing in interior wood's moisture content (not considering the effects of HVAC). The first number is the average interior moisture content in January. The second is for July. The data give a good starting point for estimating the expected interior moisture range, but modern HVAC can artificially change an environment to any condition.

(Adapted from the U.S. Department of Agriculture, Forest Products Laboratory.)

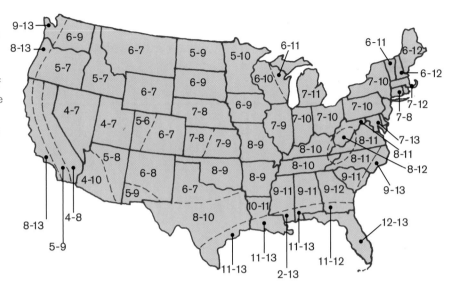

Fortifiber's Aquabar® "B" installed under the flooring will slow the migration of moisture from below.

STORE FLOORING WHERE IT WILL BE INSTALLED

Wood flooring may, or may not, be delivered at the correct moisture content to be installed in a particular house. Even the most reputable seller can only control the moisture of the flooring while it's in their possession. They can't know what happened to it in transit, or whether the mill dried it enough or too much. The only thing you can do is acclimate the flooring to the midrange of the house's moisture level. The process of allowing flooring to adjust to the moisture level of its environment prior to installation is called *acclimation*. Essentially, you acclimate wood flooring by storing it where it will be installed. Don't store wood flooring in the garage, basement, or any other place where the moisture condition differs from where the flooring will be installed.

Wood flooring should be acclimated until it reaches the correct moisture content for its environment. Measure the wood's moisture content with a moisture meter (see chapter 12). How long the acclimation process takes depends on many factors, including how much the moisture content needs to change and the species of wood. Moisture

Not all wood flooring products should be acclimated. Some engineered products come wrapped in plastic, which keeps them at the manufacturer's desired moisture content until installation. The tongue-and-groove joint in these products is precisely machined, and moisture changes can make it too tight and hard to fit.

RELATIVE HUMIDITY AND WOOD MOISTURE CONTENT

Measuring the relative humidity and temperature of the area to receive wood flooring provides a good indication of what the acclimated moisture content of wood flooring should be. The chart indicates the expected moisture content of wood flooring if acclimated under measured conditions.

TEMPERATURE		MOISTURE CONTENT (%) AT DIFFERENT RELATIVE HUMIDITY VALUES																		
°C	(°F)	5%	10%	15%	20%	25%	30%	35%	40%	45%	50%	55%	60%	65%	70%	75%	80%	85%	90%	95%
-1.1	(30)	1.4	2.6	3.7	4.6	5.5	6.3	7.1	7.9	8.7	9.5	10.4	11.3	12.4	13.5	14.9	16.5	18.5	21.0	24.3
4.4	(40)	1.4	2.6	3.7	4.6	5.5	6.3	7.1	7.9	8.7	9.5	10.4	11.3	12.3	13.5	14.9	16.5	18.5	21.0	24.3
10.0	(50)	1.4	2.6	3.6	4.6	5.5	6.3	7.1	7.9	8.7	9.5	10.3	11.2	12.4	13.4	14.8	16.4	18.4	20.9	24.3
15.6	(60)	1.3	2.5	3.6	4.6	5.4	6.2	7.0	7.8	8.6	9.4	10.2	11.1	12.1	13.3	14.6	16.2	18.2	20.7	24.1
21.1	(70)	1.3	2.5	3.5	4.5	5.4	6.2	6.9	7.7	8.5	9.2	10.1	11.0	12.0	13.1	14.4	16.0	17.9	20.5	23.9
26.7	(80)	1.3	2.4	3.5	4.4	5.3	6.1	6.8	7.6	8.3	9.1	9.9	10.8	11.7	12.9	14.2	15.7	17.7	20.2	23.6
32.2	(90)	1.2	2.3	3.4	4.3	5.1	5.9	6.7	7.4	8.1	8.9	9.7	10.5	11.5	12.6	13.9	15.4	17.3	19.8	23.3
37.8	(100)	1.2	2.3	3.3	4.2	5.0	5.8	6.5	7.2	7.9	8.7	9.5	10.3	11.2	12.3	13.6	15.1	17.0	19.5	22.9

(Adapted from the U.S. Department of Agriculture, Forest Products Laboratory.)

Narrower flooring—less than 3 in. wide— can be spread out in the room where it's to be installed to acclimate.

moves through some wood species slower than others, and wood finishes slow acclimation. If you store wood flooring in boxes or tightly stacked so air can't flow, it will acclimate very slowly.

To minimize the amount of wood movement throughout the year, the flooring should be acclimated to the approximate midpoint of the expected interior seasonal moisture range (see the map on p. 15). The data give a good starting point for estimating your expected interior moisture range, but modern HVAC can artificially change an environment to any condition. It's easy to find the interior relative humidity by using an inexpensive relative humidity meter available at many department stores.

Properly functioning central heating and air-conditioning systems control a home's environment. But these systems are often turned off during certain times of the year, subjecting wood flooring to fluctuations in moisture content, and malfunctioning HVAC systems can produce moisture levels that are excessively low or high. It's usual for the interior wood moisture content in some areas of the country to range from 6 to 12%. For a home that has a moisture range of 6 to 12% throughout the year, strip flooring should be acclimated to approximately 9%. More attention should be given to wider flooring boards since they swell and

STICKERING FLOORING BOARDS

Flooring wider than 3 in. is susceptible to cupping if one side is laid on
the floor as is done with strip flooring.
Stack it with spaces between the
boards and stickers—strips of clean,
dry wood—between the layers to
promote even airflow.

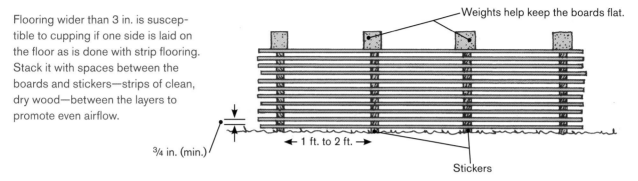

Weights help keep the boards flat.

¾ in. (min.)

← 1 ft. to 2 ft. →

Stickers

shrink more than narrower ones. I try to acclimate wide plank flooring
to slightly higher than the midpoint. It's better to allow the floor a
little more tolerance for expansion because the risk of buckling always
outweighs the possibility of having a small gap between the boards
during the dry season.

To speed up acclimation, remove flooring that's less than 3 in. wide
from its packaging and spread over a clean, dry subfloor covered with a
vapor retarder. Because of its width, plank flooring is more susceptible to
cupping than strip flooring. Stack boards wider than 3 in. in a pile with
stickers (flat strips of wood at least ¾ in. thick) between the boards to
aid uniform air circulation. Weight the top of the pile to help keep the
boards flat. To speed acclimation, use a humidifier or dehumidifier as
appropriate.

If acclimating your wood flooring sounds like a lot of trouble,
consider this. The flooring industry suffers one billion dollars a year
in losses associated with moisture problems. If you do not acclimate
the flooring correctly, you are liable for all associated financial losses. I
receive calls every week from contractors about floors that failed because
they were installed under improper moisture conditions. It's really quite
simple to escape potential problems. The key point to keep in mind is
to install the floor at the right moisture content for the environment.
Sometimes, the wood comes at the correct moisture content and needs
no acclimation. On the other hand, acclimation on some specialized
commercial projects I've done took two months. It was time well spent.

Even Drying Promotes Straight, Flat Flooring

Because wood dries and
shrinks first at the outside, the
outer surfaces tend to be in
tension while the inside is in
compression. Keeping these
forces balanced is crucial to
keeping the wood flat and
straight. That's why even airflow is so important during
acclimation. One edge drying
faster than the other can result
in crooked wood, because that
edge exerts more tension on
the board. Similarly, one face
drying faster than another can
cup or bow the lumber. Twists
result from a combination of
forces. Similar effects can
occur from uneven wetting of
the lumber.

CHAPTER TWO

Job Preparation

ALTHOUGH MOST OF THE SATISFACTION OF installing a beautiful wood floor comes from actually putting down the flooring and not from preparing for the installation, poor preparation often means that beauty will be short-lived. Whoever works on the wood floor is responsible for ensuring that all conditions are ready for the flooring project to commence. It's always wise to look for potential problems, and many potential problems are water related.

Most wood flooring failures are the result of excess moisture. It's crucial for the flooring installer to verify the moisture content of both the flooring and the structure before proceeding.

Water stains on the subfloor are a telltale sign of leaky doors and windows. Be sure doors and windows are properly flashed so as not to leak before installing a wood floor.

Preparing the Building

The first step in preventing water damage to wood flooring is to ensure that no moisture can enter the building. Inspect the exterior of the building thoroughly. Water should drain away from the building and not into it; downspouts from the gutters should also drain away from the building; and windows and doors should be installed and not leak. Moisture from any of these sources has ruined countless wood floors.

Houses under construction are invariably wetter than they will be in service. Allow construction dampness—from fresh concrete, wet lumber, plaster and drywall finishing, and drying paint—to dissipate before delivering the wood flooring. How long to wait depends on how much moisture was initially put into the building and what is being used to remove it.

Wood flooring should be delivered only once the house is at the correct moisture level (see chapter 1). Most manufacturers produce flooring with moisture content between 6 and 9%, and they recommend the flooring be maintained in this range for best performance. This coincides with the normal comfort range for humans, 30 to 50% RH at 60 to 80°F. To help maintain this moisture content in the flooring, the average moisture content of the framing members should be below 14% before delivering the flooring. Floor joists, framing bottom plates, and doorways are generally a good place to measure moisture content with a moisture meter.

The heating and cooling system should also be running before delivering the wood flooring. This will aid in removing residual construction moisture and help get the interior of the home to its normal expected environmental conditions. You may need to run temporary HVAC equipment such as portable heaters, air conditioners, blowers, and dehumidifiers on-site if the home's permanent system is not operational. It may take a week or two with the HVAC operating to remove the residual moisture from the building.

MOISTURE IN THE BASEMENT

Basements (and crawlspaces) in both existing and new buildings can be areas of high humidity. While the effect of this on flooring installed in a finished basement is obvious, moisture can migrate upward from any foundation and affect wood floors in the stories above. Moisture is liberated from the building's foundation as the concrete cures. There may also be moisture vapor permeating through the pores of the concrete from the ground under the concrete slab. Even in new construction, many concrete slabs do not have vapor barriers under them to prevent moisture migration. The ones that do often have only a piece of 6-mil plastic with holes in it, which negates much of the effectiveness of the barrier.

Throughout the summer in much of the country, basements and crawlspaces are cooler than the outside air coming into them. As the air cools, the humidity level rises—a problem exacerbated in humid areas, where the cool surface of the concrete may condense moisture out of the air. This moisture can migrate into the flooring. It's sometimes common for the subflooring and the bottom of the wood flooring above the foundation to have moisture contents ranging from 12 to 17%. This level is unacceptably high, and basements and crawlspaces may require dehumidification to remove moisture from the air.

Buildings under construction get wet—from rain before the house dries in to moisture dissipating from materials such as green lumber, concrete, plaster, and paint. Always be sure the building is dry before installing hardwood floors.

Flooring grades take into account not only defects such as knots but also length and color variation. And what's a defect to some people is character to others, as evidenced by the range of color in the hickory (left) and the knots in the Australian cypress (right).

Choosing Flooring Materials: Grades and Species

Hardwood flooring is classified by grade, species, and type. The grade generally describes the surface characteristics of the wood, lengths of the flooring, and milling tolerances. There are several grading systems for hardwood flooring. Various associations create these systems, and use of the systems by manufacturers is voluntary, so caveat emptor. Two of the more common systems are those created by the Maple Flooring Manufacturers Association (MFMA, maplefloor.org) and the National Wood Flooring Association® (NWFA, woodfloors.org). The former system applies primarily to maple; the latter is universal, but mostly used with oak flooring (see the sidebar on the facing page).

Today, there are more options for hardwood floors than ever before. Hundreds of exotic woods are available from all over the world. Each species has its own characteristics, varying in color, hardness, and dimensional stability. Selecting wood for a floor is mostly a matter of personal taste (and budget), but the environment it will be subjected to

Clear red oak.

Select red oak.

No. 1 common red oak.

No. 2 common red oak.

Hardwood Flooring Grades

All of the grades listed here will make a serviceable floor. The differences are mainly aesthetic, such as the presence of knots, sapwood, and color variations. However, wood with fewer defects tends to be more stable and predictable, so you might expect greater or more varied seasonal movement with lower grades. Of course, price is another difference, with higher grades usually costing more.

MFMA Maple Grades

First Grade. The highest standard MFMA (Maple Flooring Manufacturers Association) grade is hand-selected to minimize the natural character variations of the species.

Second Grade. The most commonly specified maple flooring; this grade exhibits more natural variations than first grade.

Third Grade. This grade has the same structural integrity as first and second grades, and exhibits more natural variation than either grade.

Third and Better. This grade is comprised of a mixture of first, second, and third grades of MFMA northern hard maple.

Utility Grade. This grade of MFMA maple may contain all defects common to maple, but the wood must be firm and serviceable.

NWFA Hardwood Grades

Clear. Clear wood is free of defects, though it may have minor imperfections.

Select. Select wood is almost clear, but contains some natural characteristics such as knots and color variations.

Common. Common wood (No. 1 and No. 2) has more natural characteristics such as knots and color variations than either clear or select grades and is often chosen because of these natural features and the character they bring to a room. No. 1 Common has a variegated appearance, light and dark colors, knots, flags, and wormholes. No. 2 Common is rustic in appearance and emphasizes all wood characteristics of the species.

There are over 100,000 different types of woods in the world. Practically every color, weight, hardness, texture, and grain pattern imaginable is available.

should also be taken into consideration. For example, light-colored floors make smaller rooms feel larger. Harder woods stand up better in high traffic areas. And areas subject to moisture fluctuation should be floored with more dimensionally stable woods.

COLOR

Wood flooring is available in a variety of colors that can change the feel of any room, making color one of the most critical aesthetic design elements. The palette of colors seems limitless. American walnut provides deep rich brown tones, hard maple has white with tan hues, and purpleheart really is purple. You can stain wood floors, but with the availability of so many colors, I generally try not to—when a stained floor is scratched, it can be hard to match the color exactly. However, I do use stain to highlight individual elements in an ornamental wood floor (see pp. 258–269).

Oxidation and sunlight change the color of wood flooring, which can be a big consideration in a room with many windows. American cherry undergoes

The color of the flooring can change the feel of a room. Dark flooring makes a large room seem more intimate, while light-colored flooring tends to enlarge small spaces.

an extreme color change, darkening to a dark reddish color within a few weeks in direct sunlight. Walnut has a medium to high degree of color change, lightening from dark brown to a golden brown. Red oak ambers slightly. Brazilian cherry starts out a tan-salmon color with some black striping and turns a rich, deep red.

HARDNESS

The relative hardness of wood species is commonly measured using the Janka Hardness Rating (see the chart at right). This test measures the force needed to embed a steel ball (0.444 in. diameter) to half its diameter into the wood being tested, with the rating measured in pounds of force. While you're not likely to be embedding steel balls in your floor any-time soon, the hardness of the wood is an important consideration.

The average woman wearing high heels can exert 2,000 lb. per sq. in. on a wood floor (and I've seen many floors severely damaged by women's heels). If you have a big dog with long nails, Eastern white pine floors with a hardness rating of 380 lb. might not be your best choice. Brazilian walnut with a hardness rating of 3,680 lb. would pass the high-heel test and give dog nails a run for their money. One last word on high heels: If a heel of one of the shoes is damaged and has a protruding nail, it can attack the floor with 8,000 lb. of force. No wood flooring can stand up to that.

The hardness of wood usually varies with the direction of the wood grain. Quartersawn flooring is a little harder than plainsawn flooring. End-grain floors were a traditional floor covering in factory buildings, heavy-traffic commercial buildings, museums, bridges, and boardwalks. They absorb energy and noise, and the angle of cut allows the growth rings to resist scraping and general wear more effectively than traditionally flatsawn or even quartersawn boards.

WOOD HARDNESS SCALE

The hardness of wood is ranked by the Janka Hardness Rating. The higher the number, the harder the wood.

WOOD	HARDNESS RATING
Brazilian walnut	3,680
Brazilian teak	3,540
Purpleheart	2,890
Brazilian cherry	2,820
Mesquite	2,345
Santos mahogany	2,200
Hickory, Pecan	1,820
Hard maple	1,450
White oak	1,360
White ash	1,320
Beech	1,300
Red oak	1,290
Birch	1,260
Heart pine (antique)	1,225
Burmese teak	1,078
Black walnut	1,010
Cherry	950
Longleaf southern pine	870
Loblolly pine	690
Douglas fir	660
Eastern white pine	380

Wood Flooring Throughout the House

People tend to expect wood flooring to be in certain parts of the house, such as halls and other high-traffic areas and formal dining rooms, but it offers advantages in most areas of a home. Hardwood flooring will outlast other materials like carpet and resilient-type flooring such as vinyl or linoleum. It also provides a comfort level that ceramic floors cannot match. Wood flooring can even work in kitchens and baths, as long as you take certain precautions: Water spills must be wiped up, bathrooms should have ventilation to remove humid air, and bath mats should be placed outside the shower. I must have seen thousands of wood floors in bathrooms and kitchens, and the only problems I have come across were associated with leaking plumbing fixtures or dripping condensation from an un-insulated toilet tank.

Wood flooring is also a good choice for bedrooms. Carpeting can be nearly impossible to keep clean of dust and pet dander, and many people with allergies find they sleep better after installing wood floors. About the only place where wood flooring is not my first choice is the foyer. I prefer stone or tile, which stands up better to the puddles of water and bits of gravel my kids track in with their snow-covered shoes.

DIMENSIONAL STABILITY

As we saw in chapter 1, wood flooring shrinks and swells as its moisture content changes, some species more than others. Plainsawn red oak flooring is the most readily available species and is often used as the baseline to compare with other species for dimensional stability. Teak is about twice as stable as plainsawn red oak. Of course, we could compensate for that by using quartersawn oak flooring, which would be almost as stable as plainsawn teak.

Types of Wood Flooring

There are three main categories of wood flooring: strip flooring, plank flooring, and parquet flooring. For the most part, all three categories are sold as ¾-in.-thick, tongue-and-grooved boards. (Engineered flooring is an exception—while it can be plank or strip, it's usually under ¾ in. thick and made up of different layers of wood glued together like plywood.) Strip flooring, the most popular category, has boards up to 3 in. wide.

Plank flooring can be considered any boards 3 in. or wider. (There is some confusion between the two, because some strip flooring manufacturers occasionally produce flooring up to 3¼ in. wide to ensure the greatest yield from the raw material supplied to the mill.) Parquet flooring is individual wooden tiles that can consist of many pieces (see chapter 7). Strip, plank, and parquet flooring are available in solid wood or engineered, and as unfinished or prefinished.

BAMBOO AND CORK FLOORING

With the growing interest in all things "green," bamboo and cork have become more popular as flooring materials in recent years. Bamboo is the fastest growing woody plant in the world and actually considered a grass. After harvesting, the bamboo is cut into strips, steamed, dried, glued, and placed in a heat press. Bamboo makes a good flooring material, but its quality varies dramatically between manufacturers.

Cork flooring comes from the bark of the cork oak tree. The cork trees are unharmed by the harvesting of the bark, and they continue producing cork for an average of 150 years. Cork flooring has millions of microscopic air pockets that give it the ability to return to its original shape after impact.

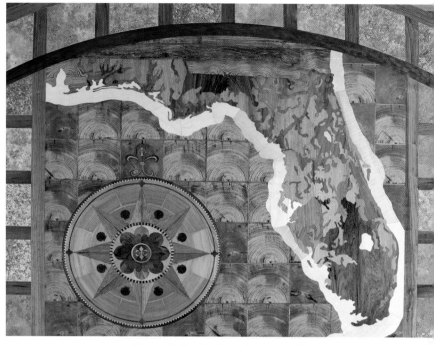

End-grain flooring is made from sections sliced from the end of thick beams. Typically glued down, end-grain flooring is extremely durable and was a favorite for old factory floors.

Estimating Flooring

Calculating the amount of wood flooring needed for a project depends on a number of factors. The first, of course, is the actual square footage of the space, but you also have to consider the shape of the room. There will be some waste on every installation, but there's usually less in rooms that are simple rectangles. Most strip or plank flooring installations in a square room require about 5% extra flooring to allow for cutting waste and any culling defects. Rooms with many angles, jogs, or bays will require 15 to 20% more wood. Both lower grades of wood and diagonal

Engineered flooring, like plywood, is composed of multiple layers with a hardwood face veneer. The multiple layers create more stable flooring.

Engineered Wood Veneers

The layers of engineered wood flooring consist of veneers laminated together at a 90° angle, which minimizes their tendency to change dimension from moisture. Engineered flooring is a great option for buildings by the water or for use in basements. Generally, the lower (or back) veneers are made from less expensive wood species. The top (or face) veneer can be manufactured from any wood species, while minimizing waste. This top layer is cut from the log by one of three different methods: rotary peeling, slicing, or sawing.

Rotary-cut veneers show a dramatic, wild graining such as that seen on plywood subflooring. These veneers provide the best yield from the raw materials but the weakest grain structure. Sliced veneers resemble sawn veneers or boards but provide better log yields. They are stronger than rotary-sliced veneers but not quite up to sawn veneers. Sawn veneer yields the least material and costs the most to produce of the three veneer manufacturing methods, but it has the strongest grain structure.

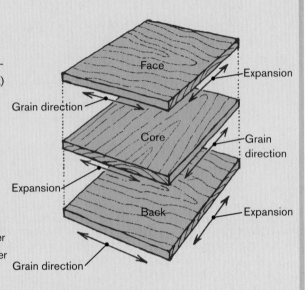

Narrow boards are known as strip flooring (above), whereas anything wider than 3 in. is called plank. Narrower boards are more dimensionally stable, but wider ones can create an old-house feel (right).

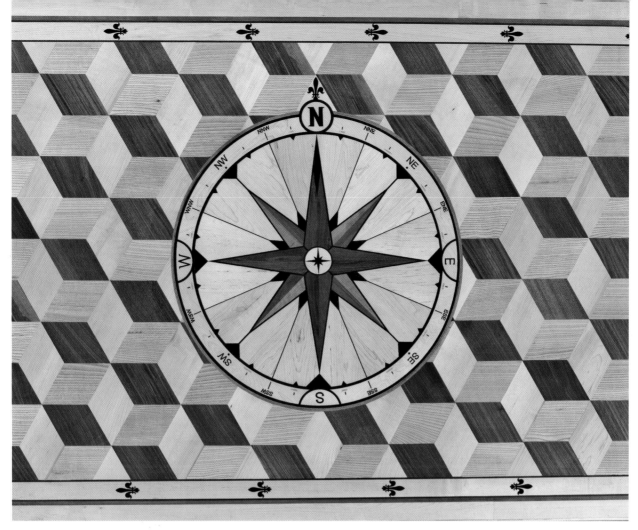

Parquet flooring is a mosaic of wood pieces combined to form a pattern. This one has a compass rose medallion at its center.

installations can add 15 to 20% waste. Lower grades may have more defects that need to be removed. And the end boards on a diagonal floor are not cut at 90°, which means that the cutoff waste has to be re-cut before they are used as starter boards on the next rows. Parquet wood flooring generally requires 10% extra material in a square room and 15 to 20% for rooms with lots of angles and corners.

National industry standards allow manufacturers to sell products with a maximum of 5% defects that do not fall within that grade of flooring. Most manufacturers of unfinished flooring abide by these standards. However, many manufacturers of prefinished wood flooring have proprietary grades for their products. A knowledgeable dealer should be able to help predict the amount of waste. Lower grades of wood will have more defects, which might lead to more waste. Then again, a defect to one person might be an aesthetically pleasing element to someone else.

Laminate Flooring

People often confuse laminate flooring with engineered wood flooring. Laminate flooring is not wood flooring. It has only a surface layer of two thin sheets of paper impregnated with melamine onto which a wood-grain effect is printed. As such, I consider laminate flooring a temporary material at best.

When estimating the amount of flooring needed, take into account the shape of the room as well as the flooring direction. Rooms with multiple jogs or bays, or diagonal installations, can result in twice the waste of simpler spaces.

Removing Existing Floor Covering

If you're installing a hardwood floor in new construction, all you have to worry about is the condition of the subfloor (this is the subject of chapter 3). But if you're working on a remodel project, you'll likely have to remove an existing floor covering, which might be carpet, linoleum, or vinyl. Removing old floor coverings is my least favorite part of installing wood flooring.

REMOVING CARPET

No matter how many times a carpet is cleaned throughout its life, removing all the animal dander, dirt, and dead bugs is just about impossible. With that in mind, I always wear a respirator when tearing up old carpet. Unlike dust masks, respirators are certified by the U.S. government to ensure that they meet specified minimum filtration requirements, as well as specific manufacturing quality levels. Also, many dust masks do not seal tightly to the face and allow airborne hazards to pass. (For more on this, see "Lung Protection" on p. 35.)

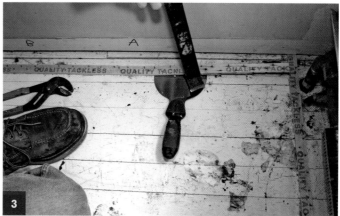

Many carpets are held in place by tack strips installed next to the walls. These narrow wooden strips have small, sharp, angled nails that point up to hold the carpet's backing. The strips themselves are nailed to the floor. To begin removing old carpet, pull back a corner with pliers. Continue around the outside of the room, releasing the rest of the carpet from the tack strip. Cut the carpet into manageable strips to haul away. If there are existing wood floors under the carpet that you simply plan to refinish, be careful not to damage them while cutting the carpet. Vacuum the subfloor as you remove the carpet strips to limit the amount of dirt that becomes airborne.

Once you've removed all the carpet, pull up the tack strips using small pry bars, a claw hammer, and pliers. A drywall-taping knife can be placed between the floor and the pry bar to protect the wood floor from damage. A cat's paw–type nail puller can also be used to remove nails from a stubborn tack strip. Wear gloves while removing the tack strips. The tack strip has sharp little teeth, and at least one will find its way into your fingers.

1. Given that you never know what sort of hazardous material can become airborne when removing old carpet, wearing a respirator is a must.

2. To remove carpet from the tack strips, grab a corner with stout pliers and pull. You can cut the carpet into manageable pieces for disposal, just be careful to avoid cutting into the floor below if you're saving it.

3. Tack strips hold carpets in place. Use a taping knife below a pry bar to avoid damage to old floors you want to save, and be careful—the tacks are sharp.

REMOVING OLD LINOLEUM OR VINYL FLOORS

You need to be careful when removing old resilient flooring. Many linoleum and vinyl floors from the 1970s and earlier contained asbestos in their backings or adhesives. Various federal, state, and local government agencies have regulations that require special licensed abatement contractors to remove material containing asbestos. (You can find these contractors online or in the Yellow Pages.) I never remove old resilient flooring without having it tested for asbestos first. It is just not worth possible health problems, breaking the law, and contaminating the home.

An alternative to removing old linoleum, vinyl, or any flooring material that possibly contains asbestos is to install a new one over it. This requires that a proper subfloor exists under the resilient floor (see chapter 3) or that new underlayment has been added. I am usually able to verify the presence of the subflooring by looking at a preexisting plumbing hole, loose flooring, or a heating vent or by removing a transition piece. The wood flooring fasteners need to penetrate the resilient-type floor and at least ⅝ in. into the actual subfloor material.

If needed, install plywood underlayment over the old resilient flooring. Most manufacturers recommend that a ⅜-in. or thicker underlayment be used. Increasing the height of the floor can sometimes present problems. For example, new wood flooring installed in an existing kitchen usually runs up to the bottom of the cabinets and in front of appliances such as dishwashers or trash compactors. This additional flooring height might make it impossible to remove such appliances for service or replacement. Also, large transitions in height to other rooms can trip an unsuspecting guest.

Many homes I work on have particleboard underlayment, a material that is unsuitable for use as subflooring, installed below the carpet. I remove it by cutting it into manageable 2-ft. squares. I find this makes it easier to pry off the floor and remove from the room without damaging the walls.

Safety Equipment Is Paramount

Installing wood floors requires an incredible amount of cutting and nailing, but the most important equipment I own doesn't cut or fasten anything. Nothing comes before safety equipment. With all the cutting,

Particleboard is considered an unacceptable substrate, so it must be removed prior to nail-down and most glue-down flooring installations. Use a circular saw with its blade set just shallower than the particleboard thickness to cut it into 2-ft. squares to ease removal.

Make Safety a Habit

Like all construction work, installing wood flooring has inherent risks. Cuts from edge tools are one obvious hazard. Others are less obvious, but perhaps more readily prevented. Reduce the risks of eye injuries, hearing damage, lung disease, and bad knees with proper safety gear. Comfort is one of the most important factors when selecting safety equipment. I tend to rely on 3M® for safety products, but you may find other brands fit you better. Try out several. Safety equipment does no good if you take it off because it's uncomfortable.

Hearing and eye protection

Respiratory protection

Kneepads

nailing, and chemicals flooring installers work with, it is only a matter of time before an injury will occur if you don't have the proper protection.

Many old-time wood flooring contractors are partially deaf because they didn't protect their hearing when using nailers and power tools. Before I started wearing proper respiratory protection, some exotic woods I used made my nose bleed. Many old-timers have severe respiratory problems. Wood dust is a carcinogen, and I have visited flooring friends who have cancer. Without safety glasses, I have had things go into my eyes, and I once had a close call with a deflected pneumatic nail. Kneepads are a very important piece of equipment. I have too many friends with permanently damaged knees. One of them almost lost his leg when a splinter in his knee became infected.

HEARING PROTECTION

One in 10 Americans has a hearing loss severe enough to affect their ability to understand speech. It's more common among wood flooring contractors. Nearly everyone I know in this trade has hearing loss, which has made me all the more adamant about hearing protection.

Wearing hearing, eye, and knee protection is an essential safety measure when installing hardwood flooring.

Noise is measured in decibels, or dB. The dB scale is logarithmic rather than linear. An increase of 10dB isn't additive; it represents a tenfold increase in noise. Consequently, even a small increase in dB can have a larger effect than is immediately apparent. Noise levels of 85dB or higher can damage your hearing. Most floor sanding equipment reaches 90dB or more. If you look on any earmuff or earplug package, you'll find a government-mandated noise-reduction rating (NRR). The NRR represents how many decibels the product reduces noise.

Because the arms of glasses interfere with how earmuffs seal to the head, wearing them diminishes earmuffs' effectiveness. To be on the safe side, it is best to wear earmuffs and earplugs together. This isn't a bad idea anyway, particularly when sanding flooring. One caveat: Use only clean ear plugs. Dirty ones can lead to infection.

EYE PROTECTION

I cannot say enough about eye protection. Without it, there is little chance that you will avoid an eye injury sometime in your career. Safety glasses have saved my eyes more times than I can count.

Protective eyewear should be made to ANSI standard Z87.1-2003, which means it should not break when smacked by a ¼-in. BB moving at 150 ft. per sec. Eyewear should also provide generous side protection for the corners of your eyes. The lenses, frames, and packaging should all be stamped Z87+ to indicate that they meet this safety standard.

Eye ware has come a long way from the goggles of years past. What's available today borders on stylish, and it's far more comfortable. Glasses with anti-fog and anti-scratch coatings are available. You can even buy prescription safety glasses through your eye doctor and from several online sources.

LUNG PROTECTION

Sanding and finishing wood flooring exposes you to fine dust particles and chemicals that attack your lungs. The American Conference of Governmental Industrial Hygienists recognizes wood dust as a human carcinogen. The size of the dust is important. Dust particles 10 microns in diameter and larger are likely trapped and expelled by the hairs and mucous of your upper respiratory system. Particles up to 2.5 microns in size settle in the lungs where they may enter the bloodstream to be filtered by the liver. Some toxic particles transport through the bloodstream to the kidneys and central nervous system. The body's immune system tries to destroy and expel toxins, but our immune system is not always successful and our cells may become cancerous if overwhelmed by toxins.

Particles that aren't expelled and don't dissolve can stay in the lungs, possibly causing allergies, respiratory problems, lung diseases, and cancer. Some finishes such as polyurethane (particularly if it's sprayed) may, over time, coat lung tissue so that it can no longer transfer oxygen to the blood. The only cure is a lung transplant.

I wear a respirator, as opposed to a nuisance dust mask. Respirators have replaceable filters that capture at least 95% of particles 0.3 microns in size or larger. They are designated by a letter followed by a number, such as N95. Respirators designated with an N are for use where there is no oil present in the air. An R designation means it's resistant to oil mists, and a P-labeled respirator is even more resistant to oils. The N-filters are most commonly used when sanding wood flooring.

Half-face or full-face respirators with the appropriate activated charcoal cartridges should be worn when applying floor finishes. Full-face respirators are best because they stop the toxic vapors from entering your bloodstream through your eyes.

KNEE PROTECTION

Working on your knees without kneepads can lead to prepatellar bursitis. Knees have a small sac called the bursa in front of the patella (kneecap). The bursa holds a small amount of fluid that allows the skin over the knee to move independently of the underlying bone. If the bursa becomes inflamed, it fills with fluid and causes swelling at the top of the knee. Without the padding provided by the bursa, kneeling would always hurt.

Charcoal cartridges for respirators are always working. Store them in a clean, sealed plastic bag or container when not in use. If you leave them out, the cartridge will be used up next time you need it.

I like kneepads with doughnut-shape pads that transfer the weight away from the kneecap. I like the shell of the kneepad to be hard enough to stop objects that might penetrate my knee. Any penetration into the fluid of the knee can cause serious infection that may require surgical cleaning to prevent loss of the leg from infection.

Moisture Meters

In my opinion, the flooring tool that ranks next in importance after safety gear is the humble moisture meter. What I said earlier bears repeating: Moisture causes most wood flooring problems. Installing wood flooring without knowing the moisture content (MC) of the building and the flooring is like driving a car blind—eventually, a crash is inevitable. The only questions are when and how bad. The relationship between wood and moisture is the most important aspect of wood science relating to wood flooring. The moisture content of wood affects its dimensions, strength, adhesive bonding, and ability to grow mold and decay fungus.

TYPES OF METERS

Moisture meters measure the electrical properties of wood—the wetter the wood, the better it conducts electricity. Indeed, once the MC of wood drops below 7%, the accuracy of most moisture meters decreases significantly. The species also affects the reading, and meters intended for use with wood should include charts to interpret the meter reading for the specific species.

The two main types of handheld moisture meters are the resistance type and dielectric type. Resistance meters measure the resistance between the two pins planted in the wood. Dielectric meters send a radio frequency field into the wood with surface electrodes and measure the resulting difference in electrical field conditions.

Although dielectric meter technology is advancing rapidly, resistance meters are currently about twice as accurate as dielectric meters. This is mainly due to the effects of varying density throughout the wood flooring. Consider dielectric meter readings qualitative and not quantitative; that is, they can indicate if there's a problem, but you can't

Reading a Moisture Meter

Meter readings may start to drift lower after driving the pins into wood flooring with high moisture content. Meter drift is less of a problem at lower moisture content levels. The best practice is to read the meter within the first two to three seconds of driving in the pins.

In dry climates, static charges may cause erratic meter readings. Minimize the effects of static by inserting the pins into the wood prior to applying power to the meter. Place your hand across the inserted pins to help discharge the static charges.

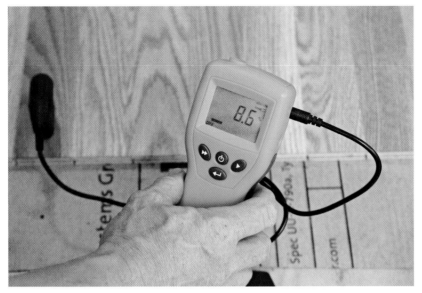

The key to avoiding most wood flooring problems is to install flooring only when the moisture levels of both the building and the flooring are within an acceptable range. A handheld moisture meter (here, a dielectric meter) is a convenient method to measure the moisture content.

trust them to tell you how severe the problem is. Their chief advantage is that they don't have to penetrate wood to generate a reading.

Moisture content can vary throughout the thickness of a piece of wood, including subflooring. When conditions are changing rapidly— say the subfloor has recently been soaked but is drying—it can be useful to know the moisture content throughout the piece. In that case, a meter with insulated electrodes driven in with a slide hammer provides the best information. In most instances, however, simply checking both sides of the wood is sufficient. And when you're checking the flooring, check several samples. If the wood is in bundles, open them up and check samples from the center of the bundle as well as the edges.

CHAPTER THREE

Subfloors: Stiff, Flat & Dry

ALL WOOD FLOORS HAVE ONE THING IN COMMON: They are only as good as the subfloor beneath them. There are a few subdivisions up the road from where I live—a couple hundred houses altogether—all built 30 to 50 years ago. I've been in many of them over the years, and most of the wood flooring is in terrible condition. There are gaps between the boards, and the floors always squeak when you walk on them. The reason? The builder installed the wood flooring over a subfloor consisting of one layer of ½-in. plywood. This subfloor isn't thick enough to hold the flooring nails, so the wood flooring moves constantly.

This story is typical. Many flooring problems have more to do with what's below the floor than with the floor itself. And while there's a lot that can go on under a wood floor, both good and bad, there's a way to work with most conditions. Solid lumber joists or I-joists could support the subfloor, or it could be one of several systems installed over a concrete slab. The subfloor itself could be boards nailed to the joists in an older home, or plywood or oriented strand board (OSB) in a more modern one. There are details you should know in each case. In the

Preparing the Subfloor

Preparing the subfloor is a crucial first step, but it doesn't require many tools.

A hammer to set protruding fasteners.

A screw gun and screws for extra holding power.

A 6-ft.- or 10-ft.-long straightedge to check the flatness of the subfloor.

A sanding machine for any high spots.

A broom to sweep the subfloor clean.

A stapler to tack down the moisture retarder.

previous example, simply using a thicker subfloor, or adding a layer of underlayment, before installing the wood flooring would have made all the difference.

Even things beyond the floor structure can affect flooring. For example, installing a wood floor over a damp crawlspace or basement without first fixing the moisture is a recipe for disaster, as the moisture can migrate into the flooring and buckle it. In the opposite way, wood floors installed over radiant heating that runs too hot can dry out and crack. Finally, there's the issue of sound control. There are simple systems that can be installed between the subfloor and the wood flooring that go a long way to prevent sound from traveling between floors.

Wood I-joists are flatter than solid lumber joists and can be engineered for greater spans. However, because they cost more than solid lumber, it's common to use the smallest size I-joists. While they're safe and code approved, such floors can feel bouncy.

Wood-Joist Floor Systems

Unless a floor is a slab on grade, it gets its support from some sort of joists. They might be solid lumber such as 2×10s or, in newer buildings, engineered lumber, commonly called I-joists. Joists made from trusses or steel are less common. Fastened directly to the joists, the subfloor creates a structural diaphragm that holds the joists together and provides an initial walking surface and a fastening layer for finish flooring such as wood. In older buildings, the subfloor is typically 1-in.-thick boards. Most new buildings have a subfloor of plywood or oriented strand board (OSB).

FLOOR FRAMING MUST BE STIFF

For wood flooring to perform satisfactorily, the floor system below must be stiff enough to support the finish floor material to prevent squeaks and sagging. An overly flexible subfloor can even cause damage to the finish along the board edges as individual pieces of wood flooring move against each other.

Stiff is not the same as strong, though the two are related. Strength refers to the ultimate load a structural member can bear before failure; stiffness is how much a member bends under load without permanent deformation. A steel cable is strong without being stiff, while a sheet of

FLOOR FRAMING IS A SYSTEM

All parts of a floor have to work together for it to function properly. Posts should be spaced so the main beam is not overspanned. Joists need to be deep enough to support their loads without bouncing excessively, and the subflooring must be thick enough to span between the joists without sagging. Mid-span bridging helps to minimize joist flexing.

Vapor retarders protect the finish flooring from moisture rising from below. Finish flooring usually runs perpendicular to the joists for better support.

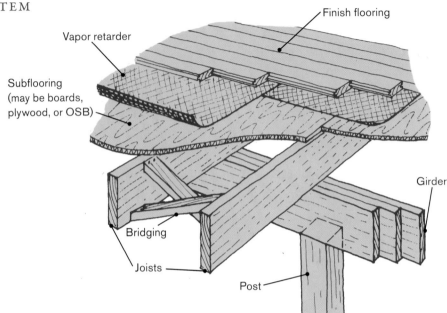

glass is stiff without being strong. Floors need to be both, but with the materials used in most construction, it's safe to assume that if the floor is stiff enough, it's also strong enough.

Stiffness is measured by the floor's deflection under load. The standard load designed for most residential floors to support is 50 lb. per sq. ft. Of that, 10 lb. or 12 lb. per sq. ft. is assumed to be the dead load, or the weight of the building materials themselves, and 40 lb. per sq. ft. is the live load, or the weight of the furniture, people, and pets the floor will support. Commercial and public building floors are designed to carry greater loads. And, of course, we've all experienced old houses whose floors were visibly sagging, and which bounced when walked on. These floors either have been weakened with time, or were inadequately designed from the start.

If you find a floor that deflects more than the standards given in the sidebar on the facing page, or that comes very close, there are several things you can do. You might be able to beef it up using one of the methods shown in the drawings on p. 44. Keep in mind that without having an engineer design the fix, you won't be able to predict the exact effect of the reinforcement. If you can't take any of the measures shown,

Standards of Deflection

The maximum deflection of a fully loaded residential floor allowed by most modern building codes is L/360. In this formula, L is the span of the joists and 360 is a factor that's been determined to result in the maximum acceptable deflection. For example, if floor joists span 14 ft., or 168 in., the most that floor would be allowed to deflect when fully loaded would be $^{15}/_{32}$ in. (168 in./360 = $^{15}/_{32}$ in.).

Building codes establish minimum standards, not best practices. To save money on framing, some new houses are designed so the floors barely satisfy the code's minimum stiffness requirement. While a floor designed to L/360 is safe, it may still feel bouncy to anyone walking across it. The china in the sideboard rattling as you walk across the dining room floor is not a mark of quality construction. Consequently, better-built homes use stiffer deflection standards, often L/480 or L/720. An L/720 floor would deflect only half what an L/360 floor does and would be far less bouncy to walk on.

There's little consensus within the wood flooring industry on floor stiffness, but it's safe to say that installing wood flooring on a bouncy subfloor can result in squeaks and finish damage, especially on prefinished products. Some experts say the deflection should be no more than L/360 for floor spans up to 15 ft. and no more than L/480 for greater spans. In any case, most designs limit the actual deflection to ½ in. It's tough to field-test for deflection, but you may be able to find the design deflection on the building plans. Generally, if a floor feels bouncy when I walk across it, I discuss stiffening it with the owner.

then you need to decide whether to install the floor. Installing the floor may actually improve the system's overall performance. However, if you do proceed, it's smart to make the owner aware in writing of the potential problems.

SUBFLOORING MATERIAL

It's common to think of the stiffness of a floor as being simply a function of the joists, but the joists work in concert with the subfloor. The subfloor helps to distribute point loads between multiple joists, and it stabilizes the joists so they don't roll under load. And whereas joists provide strength and stiffness along their length, the subfloor provides these qualities between the joists.

There are three main types of acceptable wood subflooring material: plywood, OSB, and solid boards. Until plywood became common in the 1940s and '50s, subfloors were made from solid wood boards. In the 1980s, OSB subflooring began to gain ground. Today, subflooring is made almost exclusively from plywood and OSB panels. These panels are more cost-effective to install and offer structural advantages, particularly

BEEFING UP A BOUNCY FLOOR

There are many ways to stiffen a floor system. Access from below gives you more options, but even when there is no access, you can glue and screw or nail a layer of underlayment from above. The thicker the underlayment, the more it helps. Not only does underlayment strengthen the floor, the added mass helps dampen vibrations. A layer of hardwood flooring also adds mass and offers a similar effect. And if there's access from below, gluing and nailing a layer of drywall, plywood, or OSB to the bottom offers a similar effect.

UNDERLAYMENT ABOVE OR CEILING FINISH BELOW

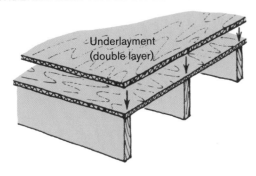

Underlayment
(double layer)

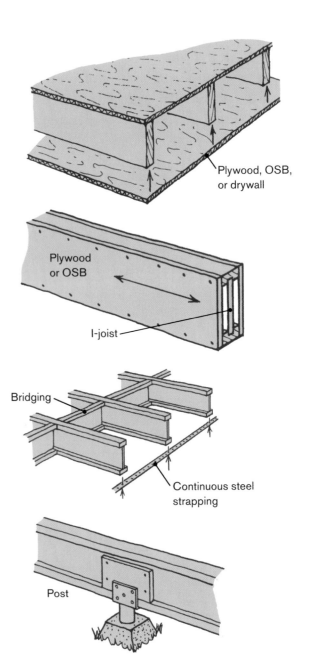

Plywood, OSB,
or drywall

SKIN BOTH SIDES OF JOISTS

With access from below, you can stiffen a floor by sistering additional joists to those already there. If they're I-joists, you can attach ⅝-in.- or ¾-in.-thick plywood or OSB to both sides of each I-joist in the bouncy area. Individual joist manufacturers should be able to recommend the nailing schedule, and using construction adhesive increases the stiffness.

Plywood
or OSB

I-joist

BRIDGING AND STRAPPING

Bridging or bottom strapping can stiffen floors in the across-the-joist direction. Particularly when combined, these approaches can reduce the vertical motion in a floor. A single row of mid-span bridging is recommended when the joists span less than 14 ft. Longer spans get two rows of bridging, each at one-third the width of the span.

Bridging

Continuous steel
strapping

BEAMS AND POSTS

With access from below, you can dramatically reduce floor vibration by adding an intermediate beam perpendicular to the joists, or by adding posts below specific joists.

Post

in seismic and high-wind zones. However, old-fashioned solid board subflooring retains wood-flooring fasteners better.

SOLID BOARD SUBFLOORING Solid board subflooring should be kiln-dried and made of so-called "Group 1" softwoods, such as northern Douglas fir, western larch, and southern pine. The boards should be at least ¾ in. thick and no wider than 6 in. (nominal). Wider boards move more during moisture changes. The boards should be installed at an angle to the joists (45° is most common). This is because it's best to install the finish floor perpendicular to the joists so you aren't relying solely on the subfloor to support the flooring between the joists. If both the board subfloor and the flooring boards were installed perpendicular to the joists, excessive movement might occur with moisture changes. Also, it would be inevitable that the edges of some flooring boards would end up directly above the joints in the subflooring, and they'd essentially be unsupported.

Even when board subflooring is in good shape, parquet flooring, thin strip flooring (less than ½ in. thick), and some engineered wood flooring can't be installed directly over it. A layer of plywood underlayment at least ⅜ in. thick is required first. It's typical for there to be some irregularity between individual subfloor boards, which in most cases can be easily bridged by ¾-in.-thick wood flooring. However, the thinner strip and engineered flooring tend to telegraph these irregularities. And because parquet flooring consists of small pieces of wood assembled together, any movement of the subfloor boards that would be accommodated by most strip or plank floors will cause gaps and squeaks in parquet. Since the plywood underlayment is more stable than board subfloors, it solves this problem.

PLYWOOD AND OSB SUBFLOORS Structural wood panels such as plywood and oriented strand board (OSB) are the most common subfloor materials today. Plywood is composed of thin sheets of veneer, or plies, laid up in cross-laminated layers to form a panel. Cross-lamination

Although plywood and OSB are the most common contemporary subfloor materials, old-school diagonally installed board sheathing holds fasteners better than either.

Most subfloors today are plywood or OSB. Plywood (above top) consists of crossing layers of veneer, while OSB (above bottom) is made by gluing together shredded lumber. Codes treat both materials equally. Both are strongest along the long dimension and should be oriented with it perpendicular to the joists.

Check the Subflooring's Label

There's a lot of information in the labels found stamped on plywood or OSB subflooring. The important things for flooring installers to look for are Exposure 1, which indicates that the panels were manufactured with exterior glue, and the thickness. You might question how important exterior glue is for an interior floor, but it does matter. Unless the house is built in a desert or during a drought, it's a safe bet it was rained on several times before the roof went on. Subfloor panels made with interior adhesives might delaminate from this amount of moisture, losing their strength. The label indicates the actual thickness, while the nominal thickness is usually $1/32$ in. thicker. This label indicates a thickness of $23/32$ in., but most carpenters would call it $3/4$ in.

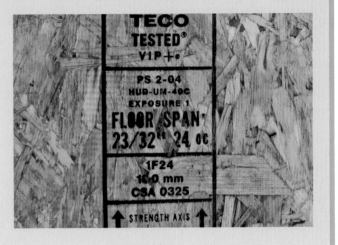

Particleboard that is currently available in the United States is not an acceptable subflooring or underlayment below wood floors. Particleboard is made of roughly ground-up wood particles glued together with a resin. It adds little strength, doesn't hold fasteners well, and provides a poor surface for adhesives.

minimizes dimensional changes, making an extremely stable product. Plywood always has an odd number of layers, with the face layers' grain usually oriented parallel to the long dimension of the panel. Plywood is stronger along its length than its width so should always be installed perpendicular to the joists.

Rather than sheets of veneer, the base material of OSB is wood that's been chipped into strands. These strands are compressed and glued together in layers (usually three to five) oriented at right angles to one another. The orientation of layers achieves the same advantages of the cross-laminated veneers in plywood. Like plywood, OSB is stronger along its length than its width. Plywood and OSB subfloor panels should be performance rated and marked to indicate that they meet national standards.

The minimum thickness of wood subfloor panels is $5/8$ in. for plywood and $3/4$ in. for OSB, both to ensure the floor is stiff enough between joists and to give the flooring fasteners enough material to obtain full holding power. The thicknesses given for wood panels are nominal. Actually, the panels usually measure $1/32$ in. thinner than nominal: A $5/8$-in. panel will measure $19/32$ in. and a $3/4$-in. panel will measure $23/32$ in.

To prevent sagging between joists, the minimum required subfloor thickness increases as the joist spacing increases. The maximum recommended joist spacing for ⅝-in. plywood is 16 in. on center. For ¾-in. plywood or OSB, the maximum joist spacing is 19.2 in. on center but the subfloor material must be tongue and groove (T&G) so that the panel edges support each other and don't sag. If the joists are spaced up to 24 in. on center, either ⅞-in.-thick T&G plywood or 1-in.-thick T&G OSB is required. Adding blocking every 2 ft. between the floor joists is required when exceeding the recommended joist spacing.

CHECK THE SUBFLOOR INSTALLATION

Before installing (or even bringing) the wood flooring into the room, I inspect the subfloor to be certain it's up to par. With board subflooring, I check to be sure the board ends are fully supported on joists and are fastened with at least two 8d ring-shank or rosin-coated nails per joist. If it's an old floor and the nails are loose, add more so the boards don't move. At the ends, the boards might need to be predrilled so new nails don't split them. Replace badly cracked boards, and set flush any fasteners that stick out of the subfloor.

If you aren't sure about the quality of a board subfloor, add a layer of plywood underlayment at least ⅜ in. thick over the subfloor.

To hold down underlayment, install nails or screws long enough to penetrate both the underlayment and the subfloor on an approximate 6-in. grid.

Although the building codes don't require glue to create a structural floor, adhering the panels to the joists with construction adhesive minimizes the chances of squeaks.

If you aren't sure about the quality of a board subfloor—for example, if it is soft pine, runs perpendicular to the joists instead of at an angle, has more cracked boards than you care to replace, or has boards that don't meet on joists—add a layer of underlayment plywood at least ⅜ in. thick over the subfloor. Plywood underlayment can reinforce a questionable subfloor and provides a smooth surface below finish flooring. Plywood at least ⅝ in. thick can also span small dips or low areas in the subfloor.

Fasten underlayment panels on a 6-in.-minimum grid pattern. While you can use ring-shanked nails or staples for this, I prefer to use screws. Whatever fastener you use has to be long enough to penetrate the subfloor fully. I also use construction adhesive to ensure the best possible bond anytime I install subfloor. Try to avoid having underlayment panel edges land on the same joists where board ends meet. In the case of underlayment over plywood or OSB, the American Plywood Association recommends at least one joist bay between panel ends in the subfloor and underlayment.

Plywood or OSB wood panels should be spaced ⅛ in. apart, except at the joints on the long edges of tongue-and-groove panels, where the panels join together to support each other. Without this spacing to

CHECKING FOR PROPER SUBFLOOR INSTALLATION

As well as being thick enough, the subfloor should be installed perpendicular to the joists, with its ends spaced by ⅛ in. to allow for expansion. The end joints should be staggered, and if the sheets are square-edged as opposed to tongue and groove, there must be blocking between the joists below the joints. Nail sheets at least every 6 in. along continuously supported edges, and at least every 12 in. in the field.

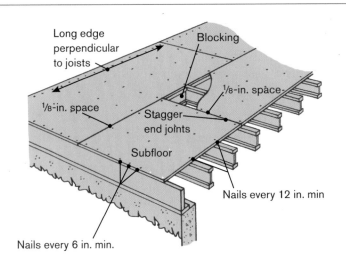

Fix Squeaks Before Installing the Finish Floor

Walk around the room listening for squeaks, and check that the nailing schedule looks right. You might have to add additional fasteners. For ¾-in. subfloor material, 8d ring-shank nails or 2-in. deck screws are a good choice. A sagging joist may leave a gap between it and the subflooring. Glue shims in such gaps, assuming of course, you have access from below (photo at near right). Squeaks coming from between joists may be caused by inadequate bridging. Nail any loose subflooring and, if necessary, install a reinforcing piece of bridging, gluing it to the subfloor (photo at far right).

provide some room for movement, excessive moisture can buckle the edge of the panels. And if the panels are in contact with each other, any movement can cause squeaks.

Some manufacturers take into consideration the ⅛-in. required clearance and size their panels appropriately smaller. Many, if not most, do not. And if the carpenters framing the building space such panels appropriately, it won't take many before the panel ends start falling past the joists. The alternative is to trim full-size panels in the field, but this is rarely done. When you encounter tightly butted subfloor panels, you can cut a space between them using a circular saw. Set its depth of cut to equal the panel thickness, and saw down the middle of the end joints between the panels. (Remember: Cut only the end joints of T&G panels, not along their length.)

Plywood and OSB subfloors should be nailed or screwed every 6 in. where panel edges run along joists and every 12 in. on intermediate joists. It is good practice to glue down the subfloor with a ¼-in.-wide bead of construction adhesive on each joist.

Protruding fasteners should be reset below the top surface of the subfloor, and high spots, particularly where panel edges have swollen from construction moisture, get knocked down with a floor sander.

Installing hardwood on subfloors with hills and valleys is a recipe for failure. Manufacturers make liquid fillers for low spots, but gluing down wood shims or plywood with construction adhesive is a cheap, foolproof solution.

KEEP THE SUBFLOOR FLAT

As well as being stiff, subfloors must be reasonably flat or areas of the wood floor would not consistently contact the subfloor, causing squeaks if fasteners don't suck the flooring down tight or adhesives aren't able to bridge the gaps. For wood flooring that requires fasteners 1½ in. or longer, the subfloor must be flat within ³⁄₁₆ in. over 6 ft. This is approximately the thickness of two quarters and a penny together. Wood flooring installed using shorter fasteners or adhesive requires a subfloor that is flat within ⅛ in. over 6 ft. (a penny and nickel). Flatness refers to the highs and lows of a floor, whether or not it's level.

While there are self-leveling liquids that set into a solid for filling low spots in wood floors, I prefer simply to glue down cedar shims. If the dip is deep enough and wider than a pair of shims laid end to end, I'll use the preferred method of filling the center with a piece of plywood and feather the edges out using shims. Use plywood whenever possible. And sand down high spots with a floor sander and coarse paper.

Once you're confident that the subfloor is sound and level, there are still a couple chores left prior to installing any wood flooring. First, if the room has its doors and moldings installed, you'll probably need to remove some and cut others back. You'll also have to sweep up. Dirt and woodchips on the subfloor can make installing the flooring difficult, and lead to squeaks and even failure with glue-down floors. Finally, you'll need to install a vapor retarder atop the subfloor.

Accommodating Existing Trim

Many times the existing trim in a room will need to be removed or cut back to facilitate the installation of the wood flooring. Using the right techniques prevents most damage and allows reinstallation of the original trim.

1. Caulk is often used to fill any cracks between the top of the base molding and the drywall. To avoid tearing the paper on the drywall when removing the base, first cut the caulk with a knife.

2. Gently insert a pry bar to open the gap between the molding and the wall.

3. Once there's a gap, proceed with a larger flat bar. Use a block of wood (or a wide taping blade) behind the bar to prevent damaging the wall.

4. Finish removing the base by hand. Use pliers to draw the nails through the base from behind.

5. You'll need to shorten the doorjambs and trim to accommodate the new flooring. The cuts can be made with a thin handsaw, such as a Japanese-style pull saw, but a better method is to use a Fein® SuperCut saw. Place a board the same thickness as the flooring under the saw blade as a guide.

Classes of Vapor Retarders

Although the common term is vapor barrier, vapor "retarder" is more accurate. Many so-called vapor barriers allow some vapor passage. This can be a good thing, as a vapor retarder can slow the passage of moisture during transient periods of high humidity, while allowing drying to take place through the retarder the rest of the time. The ability for a material to slow the passage of moisture is rated in perms, with lower numbers allowing less moisture movement. For use on wooden joist subfloor systems, vapor retarders with perm ratings between 0.7 and 50 are recommended.

- Vapor impermeable: 0.1 perms or less
- Vapor semi-impermeable: 0.1 to 1.0 perms
- Vapor semi-permeable: 1.0 to 10 perms
- Vapor permeable: greater than 10 perms

The vapor permeability of old-school materials is off the charts or unknown. Better is to use vapor retarders designed specifically for wood flooring, such as Fortifiber's Aquabar "B."

VAPOR RETARDERS KEEP MOISTURE OUT

As discussed in chapter 1, environmental moisture is the leading cause of flooring problems. One way this happens is when water vapor travels by diffusion from basements or crawlspaces through the subfloor into the finish flooring materials. Diffusion is when molecules move from an area of high concentration to an area of lower concentration. Water vapor travels from areas of higher relative humidity (or higher vapor pressure) to areas of lower humidity. The greater the vapor pressure difference, the more moisture vapor will diffuse in a given time.

The rate of diffusion through a material depends both on the degree of vapor pressure that's pushing the moisture and the material's permeability. Permeability is a measure of the amount of water vapor that can pass through a specified material in a certain amount of time. The degree of permeability is expressed in perms. Materials with high perm values allow more moisture to pass through than those with lower perm values do. (The perm rating of vapor retarders is measured using ASTM Method E96, which measures water vapor transmission in a specific controlled environment.)

Vapor retarders slow the upward movement of moisture, helping to prevent occasional elevated moisture conditions below from buckling, cupping, or cracking the flooring. Vapor retarders usually go between the finish floor and the subfloor, except when you're installing a wood subfloor over a concrete slab. Then, the vapor retarder goes between the concrete and the subfloor. Vapor retarders can come on a roll such as tarpaper does, or they can be trowel-applied liquids. The National Wood Flooring Association recommends a perm rating of no less than 0.7 and no more than 50 for vapor retarders used between wood flooring and wooden subflooring. Vapor retarders with too low a perm rating can cause problems by stopping moisture and allowing it to condense on the subfloor. For example, when installing 6-mil polyethylene film with a perm rating of 0.06 over ¾-in. OSB (which has a perm rating of 0.49), moisture could be trapped between the plastic sheet and the top of the wooden subfloor causing possible mold or rot. On the other hand, red rosin paper has a perm rating over 100 and provides no moisture protection.

Commonly installed on roofs and walls, #15 or #30 asphalt-saturated building paper (also called tarpaper or builder's felt) is frequently used under wood flooring as a vapor retarder. These products tend to meet ASTM Standard D-4869, but the standard does not specify a required perm rating like Federal Specification UU-B-790. This fact makes me nervous. I do not like taking chances on possible moisture problems. Instead, I use specialized vapor retarders specifically designed for wood floors. For most wood floors, I use Aquabar "B" (Fortifiber® Building Systems Group, www.fortifiber.com). Aquabar "B" meets Federal Specification UU-B-790 and has a specified perm rating of 0.87.

For glue-down applications or over radiant heating, I use MVP4 urethane membrane (Bostik®, Inc., www.bostik-us.com). MVP4 is a trowel-applied urethane membrane designed to reduce moisture vapor transmission from the subfloor. It also creates a noise-reduction barrier and acts as an anti-fracture membrane, an important characteristic when installing over a concrete slab. It allows the membrane to bridge cracks (up to ⅛ in.) that would allow moisture to move up through the slab. MVP4 is applied using the notched side of a ³⁄₁₆-in. × ⁵⁄₃₂-in. V-notch trowel. Properly troweled, it levels out to a flat, uniform 30-mil membrane (about the thickness of a credit card).

Start laying the vapor retarder along the same wall as the flooring installation, and lap all seams by 4 in. or more. By starting at the same wall as flooring installation, the flooring won't catch and tear the vapor retarder during installation. Staple the vapor retarder to the subfloor to hold it in place during the flooring installation.

To minimize the potential for radiant-floor heating to cause the release of VOCs from asphalt-based moisture retarders, the author uses a urethane vapor retarder. Applied with the specified V-notch trowel, the product self-levels to about the thickness of a credit card.

Concrete Floors Need Leveling, Too

No matter which method you use to install wood flooring on a concrete slab, the slab must be flat. First, clean the concrete of loose materials, oil, grease, sealers, curing compounds, waxes, and any other surface contaminants that may inhibit bonding. A floor buffer with 12-grit or 24-grit paper will clean the surface quickly. Then, fill the low spots with a self-leveling, cement-based underlayment rated at least 3,000 psi. Spread the leveler using a metal straightedge, finish trowel, squeegee, or similar tool. Take down the high spots using a grinding machine fitted with abrasive stones. Both the buffer and the grinder can be rented.

Lower-strength concrete (under 3,000 psi) should not have wood flooring glued directly to it, because it may not be as strong as the adhesive and will be the weak link in the system. The current IRC minimum for residential slabs is 2,500 psi, which is what most builders use. Older slabs might not meet this standard. A nail drawn across lightweight concrete will generally leave a scratch. Use a floating wood floor or floating subfloor in this situation.

Preparing a Concrete Slab for Wood Flooring

Concrete slabs are a common substrate in single-story buildings, and on the first floor of multiple-story ones. In much of the south and west, slab-on-grade construction is standard. In the north, most homes have basements and owners often want to finish them. Concrete slabs can bring their own problems to flooring installation. First, because they're installed on the ground, unless detailed properly they can be a direct conduit for moisture. And unless you're there to see the slab installed, there's no way to know for sure if it's properly detailed. And unless expertly placed and finished, concrete slabs are notorious for being out of flat. Finally, installing a wood floor over a concrete slab may require building some sort of wooden subfloor assembly first.

There are a number of ways to install wood floors over concrete (including direct glue-down, floating subfloors, and the screed method), but in all cases, the slab should be clean, flat, dry, and structurally sound. The concrete should be flat within $\frac{3}{16}$ in. over 10 ft., or $\frac{1}{8}$ in. over

6 ft. If it's not, the remedy is to apply a self-leveling compound such as Bostik SL 150 high-compressive strength, self-leveling, cement-based underlayment to level low spots, and to grind down the high spots.

INSTALL THE VAPOR RETARDER FIRST

Concrete slabs are like large rock sponges. They have the potential to supply large amounts of moisture that can damage wood flooring. Be sure to perform proper moisture testing (see chapter 12), although testing indicates only that the concrete is within the correct moisture level range *at that particular time*. Because of concrete's potential to supply a tremendous amount of moisture to wood flooring, I use a vapor retarder with a lower perm rating than I do over wood subfloor. Installing a vapor retarder with a perm rating 0.13 or lower between the concrete and wood flooring is imperative to avoid future problems.

There are three common ways to do this. You can use a minimum 6-mil construction-grade polyethylene film. The plastic should meet ASTM D-1745 (the information should be on the package) and have high tensile, tear, and puncture resistance. In the past, it was common to apply two layers of #15 asphalt-saturated paper, adhering both layers with an appropriate adhesive (usually asphalt mastic). Because of concerns about VOCs, this method is now rarely used. Normal sheet vinyl may also be an effective vapor retarder, but you cannot use cheap adhesive to install it (cheap adhesive generally fails when subjected to moisture). Finally, a urethane membrane such as Bostik MVP4, or other chemical system accepted by the flooring manufacturer, can give good results. I prefer this method.

DIRECT GLUE-DOWN

Once the slab is level and a vapor retarder such as MVP4 urethane applied, it's possible to lay some wood flooring directly on top of concrete. Strip flooring under ½ in. thick, solid parquet flooring, and engineered wood flooring can be glued to concrete, though it can be a challenge to force and hold the pieces of solid wood flooring together while the adhesive cures.

Even though some adhesive manufacturers warrant it, I don't recommend solid (¾ -in.-thick) strip flooring and plank flooring for

Humidity affects the cure rate of moisture-cured adhesives to a greater degree than temperature—the higher the humidity, the faster the cure. Under normal conditions, light foot traffic is acceptable after 8 to 10 hours and normal traffic after 24 hours.

When direct-gluing wood flooring to concrete, the author always trowels on a urethane vapor retarder first and then uses a compatible glue. Holding down the boards while the glue sets can be a challenge, so it's a good idea to use a few strategically placed mechanical fasteners as well.

direct glue-down to concrete. Many boards are not perfectly straight and the tongue and grooves don't fit tightly. This is relatively easy to work with when you're using a flooring nailer that both helps to force the boards straight and then mechanically fastens them to the subfloor. However, there are no mechanical fasteners used when directly gluing flooring to concrete, and it's extremely difficult to draw imperfect flooring tightly together and then hold it there while the glue sets. However, the practice is becoming more common in order to save money when installing over concrete. Gluing flooring directly to the concrete slab is the most economical method.

In addition to direct glue-down, there are several other ways to install wood floors over concrete. They all involve creating some sort of framing or subfloor between the concrete and the finish flooring. The most cost-effective of these methods is to create a floating subfloor composed of two layers of plywood. All other methods require the expense of gluing or fastening the subfloor and any framing to the concrete slab.

SCREED METHOD

Screeds are generally pieces of 2×4 nailed, screwed, or glued to the concrete slab. Since it is difficult to obtain perfectly straight lumber, pieces shorter than 48 in. are used to be sure they conform to the concrete. Standard 3/4-in. strip flooring can be installed directly over screeds, and thinner engineered flooring can be installed over screeds if there's a 3/4-in. plywood subfloor installed first.

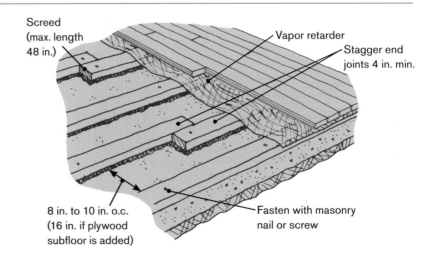

Screed (max. length 48 in.)

Vapor retarder

Stagger end joints 4 in. min.

8 in. to 10 in. o.c. (16 in. if plywood subfloor is added)

Fasten with masonry nail or screw

USING SCREEDS OR SLEEPERS

Nailing hardwood flooring directly to screeds (or "sleepers") installed over a vapor retarder is a common method of installing over concrete slabs. Screeds or sleepers are usually made from kiln-dried, pressure-treated 2×4s laid flat and fastened to the concrete with masonry nails or screws. Tongue-and-groove solid board flooring at least ¾ in. thick and less than 4 in. wide may be installed directly over screeds spaced 8 in. to 10 in. on center. No subfloor is needed, though an alternative is to space the screeds at 16 in. and add a layer of ¾-in. subfloor. Because boards are typically less than ¾ in. thick, engineered wood flooring generally is not recommended for installation directly over screeds.

For many years it was common practice to lay screeds directly on their flat face into hot (poured) or cold (cutback) asphalt mastic. The end joints were staggered so the ends lapped at least 4 in. As you might imagine, the smell of the asphalt drove this method from favor, and urethane adhesives have been substituted instead.

I generally do not use screeds. They're more expensive and labor-intensive, and adhesives have improved so much that most floors are directly glued to the concrete or installed on a floating plywood subfloor. (Today, screeds mainly reside under athletic floors.) About the only time I use them is if the concrete slab is in such bad shape that other methods won't work, or if insulation is required under the floor.

PLYWOOD SUBFLOOR ON SLAB

Plywood nailed or screwed directly to a concrete slab is an economical alternative to screeds. Additional layers may be necessary to accommodate the length of fasteners.

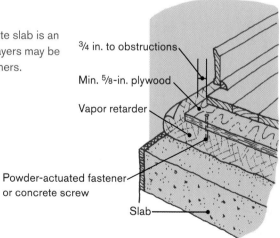

¾ in. to obstructions

Min. ⅝-in. plywood

Vapor retarder

Powder-actuated fastener or concrete screw

Slab

Concrete Fasteners

Although hammer-driven hardened nails will do the trick, it takes a lot of time to fasten anything this way. Better to use powder-actuated fasteners, which drive nails into the concrete with a gunpowder charge. Like hand-driven masonry nails, powder-actuated fasteners rely on compression from the concrete that the nail displaces for their holding power. Additionally, friction heat generated by the powder-actuated fastener entering the concrete creates what's called a sintering effect. In essence, it creates a partial chemical bond between the concrete and fastener. Generally, 1-in. penetration into the concrete is sufficient.

Concrete screws are installed into holes drilled into the concrete with a hammer drill. The holes are slightly smaller in diameter than the screw, and about ¼ in. deeper to prevent the screw from bottoming out. Manufacturers usually provide the proper-size drill bit with each pack of screws. Concrete screws generally require a minimum embedment of 1 in. and should be embedded no more than 1¾ in. Screws are available from Buildex® (tapcon.com), while Hilti® (hilti.com) makes guns, charges, and fasteners.

The fastest way to attach wood to concrete is to use powder-actuated fasteners. They're driven by a gunpowder charge, and the tool requires the same respect as a firearm: Hearing protection and safety glasses are a must.

MECHANICALLY FASTENED PLYWOOD SUBFLOOR

One way to install a plywood subfloor over concrete is simply to nail or screw it into place (see the drawing on p. 57). The minimum recommended plywood thickness is ⅝ in., and it must be thick enough to accept the length of fastener required by the wood flooring being installed. The plywood needs a designation of Exposure 1 to ensure it's manufactured with exterior adhesives. If pressure-treated plywood is used, it must be kiln-dried and not have an elevated moisture content (check with a moisture meter if you're unsure).

Fasten the plywood with either powder-actuated fasteners or concrete screws every 6 in. along the edges and every 12 in. in the field (at a minimum). And fasten the plywood working out from the center to ensure it will lie flat. Stagger the sheets of plywood, spaced ⅛ in. from other sheets and ¾ in. from vertical obstructions.

GLUED-DOWN PLYWOOD SUBFLOOR

Gluing plywood to a concrete slab creates an incredibly strong bond and minimizes the number of holes through the vapor retarder. The minimum recommended thickness for the plywood is ⅝ in., and it should be designated as Exposure 1. You may need to use thicker plywood or multiple sheets so it's thick enough to accommodate the flooring fasteners

LEFT Directly gluing plywood to a slab coated with a trowel-applied vapor retarder is fast and creates a strong bond. Scoring the bottom of the sheet with a circular saw helps it conform to minor slab irregularities.

RIGHT The glued-down sheets of plywood should be spaced at least ¾ in. from any vertical surfaces.

you intend to use. The plywood is generally cut in half, either lengthwise or in width, and its bottom is scored in about a 1-ft. grid pattern using a circular saw adjusted to cut to approximately half the plywood's thickness. These measures ensure the plywood is flexible enough to conform to minor irregularities in the slab.

Stagger the sheets of plywood, spaced ⅛ in. from other sheets and ¾ in. from vertical obstructions. I use Bostik's® Best moisture cure urethane adhesive (see Appendix C on pp. 326–328) for all such installations, applying the adhesive over Bostik's MVP moisture retarder. Holding the plywood down firmly enough to make full contact with the adhesive while it cures is the biggest challenge. You can weigh it down (I often stack the hardwood flooring on the plywood) or shoot in concrete fasteners at strategic locations.

FLOATING WOOD SUBFLOOR

A floating wood subfloor, which is the preferred method used by most wood flooring professionals, is made from two layers of minimum ⅜-in., Exposure 1 plywood. Floating floors are not fastened to the concrete, so there are no holes in the vapor retarder and no glue to mess around with. The first layer lies in line with the walls. Stagger both layers of plywood, spaced ⅛ in. from other sheets and ¾ in. from vertical obstructions. Lay the second layer at an angle (45 or 90°) to the first layer. (Plywood is generally only placed on a 45° angle when it needs to span multiple rooms. Installing on an angle helps to avoid joints in the doorways, which might lead to problems with the finish floor.) Fasten the layers together on a 12-in. grid pattern in the field and every 6 in. around the perimeter. My floating subfloors generally consist of two layers of ½-in. plywood glued with construction adhesive and screwed together.

FLOATING WOOD SUBFLOOR

Two layers of plywood comprise a floating subfloor. The layers are fastened only to each other, and "float" above the slab. This method is the author's preferred installation over concrete.

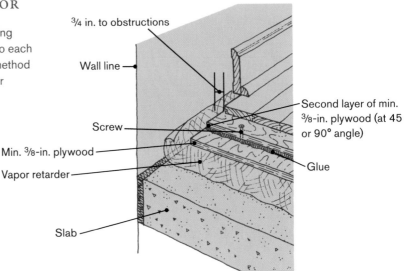

¾ in. to obstructions

Wall line

Screw

Min. ⅜-in. plywood

Vapor retarder

Slab

Second layer of min. ⅜-in. plywood (at 45 or 90° angle)

Glue

Controlling Sound

Many homeowners value sound control between floors; it's also a code requirement in multifamily buildings. Particularly in multifamily applications, it is critical to know the standards required for the local building codes and the design specifications of the building. There are two main types of sound that building designers concern themselves with: *impact sound*, which is generated by footfalls or objects coming in contact with the flooring, and *airborne sound*, which comes from voices or sound systems.

Most building codes require floor/ceiling assemblies in multifamily housing to transmit a reduced level of impact and airborne sound. Generally, codes call for a minimum 50dB sound reduction between dwelling units, referencing standards called Sound Transmission Class (STC) 50 and Impact Insulation Class (IIC) 50. (It's more complicated than that on a scientific level, but for these purposes 50dB is fine.)

WOOD FLOORS OVER ACOUSTIC UNDERLAYMENT

ENGINEERED WOOD FLOORS WORK IN ONE LAYER ON ACOUSTIC UNDERLAYMENT

After rolling out the acoustic mat underlayment, engineered floating floors can be installed with no further prep. Be sure to install the perimeter isolation barrier, which is usually the same material as the acoustic underlayment.

FLOATING PLYWOOD FLOOR ALLOWS NAIL-DOWN INSTALLATION ON ACOUSTIC UNDERLAYMENT

If floor nails penetrate the acoustic underlayment, they'll carry sound. By creating a two-layer floating floor above the underlayment, it's possible to use traditional nail-down flooring.

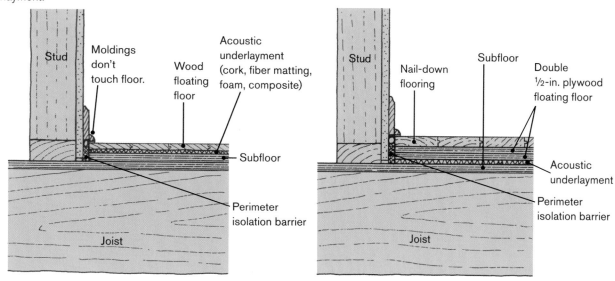

Those standards can be met one of two ways. Generally, the designer specifies wall and floor/ceiling assemblies proven by ASTM testing, and the builders and installers follow the directions. Manufacturers of flooring underlayment sell tested material systems that promise certain levels of noise abatement, and these can be worked into the design specifications. Alternatively, the building can be field-tested for compliance after construction. In that case, the referenced standards require only a nominal 45dB reduction (STC 45 and IIC 45). Sound testing is beyond the scope of this book, so I'll focus on designer-specified methods.

Wood flooring doesn't interact with the impact and airborne sound to the same degree. Different floor coverings and underlayment systems on a given subfloor assembly have not been shown to make an appreciable difference in airborne sound transmission, so the STC standard isn't very useful for flooring products. Consequently, the IIC standard is the accepted tool for comparing wood flooring and underlayment products on a given subfloor assembly.

UNDERLAYMENT SYSTEMS DAMPEN SOUND

Impact sound travels through rigid structures with little loss of energy. A good example is how someone tapping on heating pipes can be heard throughout an entire building. Impact sound transmission can be controlled by isolation and absorption. It's a common misconception that the mass of the flooring structure plays a big role in sound attenuation. The fact is that a concrete slab floor with a density of 100 lb./sq. ft. is only slightly more effective in retarding impact sound than wood frame construction with a density of 10 lb./sq. ft.

There are a variety of methods and products available to reduce sound transmission through floors. Many sell as systems, and the specifications differ between them. The important point is to isolate the wood flooring from the building structure. Common products for sound control consist of an underlayment of cork, fiber matting, foam, or a composite membrane. Solid wood flooring can be direct-glued to acoustical material that is glued to the subfloor (see the drawing on p. 61). Engineered wood flooring can be floated over a layer of acoustic material such as a foam or cork pad meant for that purpose. Strip or plank flooring can be nailed to a plywood subfloor that's floated over a layer of acoustic material, as long as the fasteners don't penetrate through the acoustic material and into the building's structure. If that happens, vibrations will carry through the fasteners to the structure and the living spaces below.

The perimeter is probably the place where people make the most mistakes. If impact sound transmits from the floor to the walls, you lose much of the effectiveness of the system. One basic key to peak performance is to avoid hard surface transference points. This means that the subfloor and floor should not contact the wall. The gap between the flooring system and the walls is generally filled with the same vibration absorbing-material as the underlayment. A small gap should be left between the base and shoe moldings and the floor. Leaving a gap prevents sound from traveling across the floor to the wall or molding and down behind the wall where there is no sound control.

WOOD FLOORS OVER RADIANT SLABS

Radiant slabs are similar to any concrete slab. Wood flooring can be installed directly on sleepers cast in the radiant slab, which makes it easy to know where the nails are going. You can also install a plywood subfloor over the sleepers first, but you have to pay close attention to nailing afterward. This is a better solution for glue-down floors.

DIRECT-NAIL TO SLEEPERS

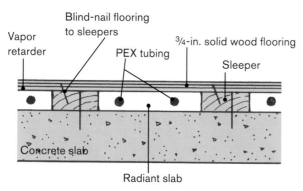

PLYWOOD SUBFLOOR OVER SLEEPERS

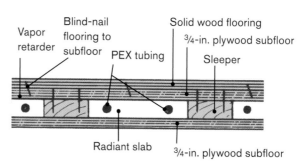

Radiant Heating Under Wood Floors

Under-floor radiant heating is becoming increasingly popular. Pipes, often of a flexible plastic called PEX, are hidden in the subfloor system. Warm water running through them heats the flooring, which in turn heats the room. Because the surface area of a floor is so large relative to that of radiators, its temperature doesn't have to be very high to heat the space. The PEX can be installed in a masonry mix poured on top of the subfloor, or it can be fastened to the subfloor from below (see the drawing on the facing page). Sometimes, radiant heat is installed in masonry slabs, and wood flooring can be installed over it there as well.

Extra care is necessary with the design and installation of both the heating system and the wood flooring. It's crucial for the heating system designer to limit the maximum surface temperature of the subfloor to 85°F, as under-floor radiant heating can dry wood flooring excessively. Sharp temperature fluctuations should be avoided, as they dramatically exacerbate moisture fluctuations in the wood flooring, which in turn repeatedly changes dimension. These repeated dimension changes stress the fasteners and can eventually lead to squeaky floors and wide joints between the flooring boards. To limit these dimensional changes, design the radiant heating system to deliver gradual temperature changes. It should incorporate an outdoor thermostat that allows the radiant system to anticipate interior heating demands. This results in more gradual floor temperature variations in response to changes to the outside temperature.

The radiant heating system should be running for at least five days prior to the wood flooring arriving on the site. It is crucial that the system be turned on even in the middle of a summer heat wave. The obvious reason for this is to check for leaks, but doing so also drives construction moisture out of the subflooring. If the subflooring is not dried out before the wood flooring is installed, turning on the heat will drive any moisture out of the subfloor and into the flooring, which will cup and buckle. I have been on many inspections where this was the problem and where the damage could not be repaired.

It is important to monitor the moisture content of the subfloor during the initial five days of the systems operation, and flooring

RADIANT HEAT ABOVE OR BELOW THE SUBFLOOR

Tubing installed above the subfloor and between sleepers is easy to avoid with nails. Installing the tubing below the floor requires shorter fasteners or great care, but adds nothing to the floor height.

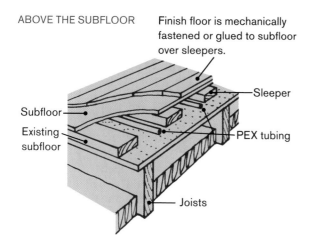

ABOVE THE SUBFLOOR — Finish floor is mechanically fastened or glued to subfloor over sleepers.

Subfloor

Existing subfloor

Sleeper

PEX tubing

Joists

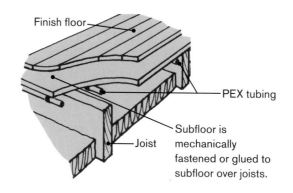

BELOW THE SUBFLOOR

Finish floor

PEX tubing

Joist

Subfloor is mechanically fastened or glued to subfloor over joists.

installation shouldn't begin until reaching the target moisture content. The ideal is to install the flooring at the average moisture content for the location over subflooring that's at the same moisture content or lower.

Both fasteners and moisture-retarding products should be compatible with radiant heating. The fasteners should be short enough that they won't penetrate heating pipes. Avoid moisture retarders and adhesives that contain asphalt as they may off-gas an asphalt smell. For glue-down floors, I use Bostik's Best urethane. On radiant installations, I use Bostik's MVP moisture retarder on both my glue-down and mechanically fastened floors.

> Acclimating wood flooring to too low a moisture content is a common mistake. When seasonal humidity rises, so will the moisture content of the wood flooring. In much of the country, the seasonal humidity is at its highest in the summer when the radiant heating system isn't operating. In these conditions, flooring picks up moisture and swells. The result will be at least cupping and maybe buckling from the stress of expansion.

CHAPTER FOUR

Installing Strip & Plank Flooring

THIS CHAPTER EXPLAINS THE BASIC LAYOUT and installation of strip and plank flooring, both solid and engineered. Later chapters will get into designs that are more ornate. Here, I'll focus on the essentials: How to lay out a good-looking floor, where to start the layout, how to keep the boards straight, and what to do when the room isn't straight. I'll also provide the basics on how to fasten flooring boards so they stay put.

While the layout of the various types of flooring is similar, the methods of attachment can differ. Strip flooring is usually fastened by blind-nailing through the tongue, though it can sometimes be glued.

Don't assume that a room's walls are parallel or square. Measure at both ends of the room, so you can plan the flooring layout to minimize the visual effect of such flaws.

Because of their greater width, planks may require more than blind nails. This can mean glue, screws with holes countersunk and plugged, or decorative cut nails through the face of the plank. Most engineered flooring is glued down.

Layout and Design

For a basic strip or plank floor, it might seem that all you need to do is find the longest wall and start there. Sometimes it's that simple, but houses are not perfect—rooms can taper, be out of square, or otherwise demonstrate the foibles of their builders. While the flooring installation can't correct these gross problems, it can often make them more or less visible. Before beginning layout or installation, I measure the room to check square and parallel walls, and generally consider how the floor will affect the look of the room. A little bit of thought can avoid many problems.

DECIDE THE FOCAL POINT BEFORE LAYOUT

In addition to the room's geometry, its main focal point is an important layout consideration. When you enter, what is the first thing that catches your eye? Focal points can be anything from a fireplace to a window that frames a view of the outdoors. Rooms can have multiple focal points, but one will be primary.

Since most peoples' eyes tend to veer toward the focal point of the room, this is where to take the most care with installation. The wood flooring should be installed to complement the focal points of the room, and the main focal point is usually the best place to start layout and installation. By starting at the focal point, you can avoid placing any boards that may need to be tapered, notched, or noticeably narrower in this highly visible location. Instead, make these adjustments along a distant wall where they'll be hidden with trim or furniture. Flooring can also accentuate the focal point of the room with a decorative border.

The focal point is a room's dominant feature. Laying out the floor with the focal point in mind enhances the overall look and minimizes the effect of compromises or adjustments made in less visible spots.

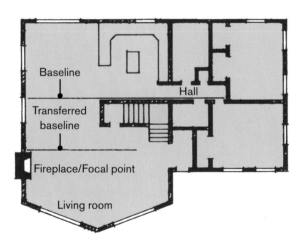

Baseline

Hall

Transferred baseline

Fireplace/Focal point

Living room

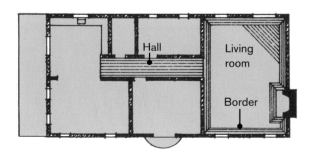

Hall

Living room

Border

FOCAL POINTS CAN DICTATE LAYOUT BEYOND INDIVIDUAL ROOMS

In the top drawing, starting the flooring in line with the long, straight wall of the hall might seem the natural place to begin. However, doing so is likely to require notching or ripping a board next to the focal point, which is the fireplace. To keep the flooring parallel to the hall and aligned with the fireplace, extend a line from the hall across the living room. Use measurements from this line to the fireplace to create a transferred baseline. Starting the floor on this line aligns it perfectly with the fireplace. If boards must be ripped, they'll be in the hallway, a less noticeable spot.

BORDERS ISOLATE PROBLEMS

In the bottom drawing, rather than running the flooring from the hall into the living room, a better choice would be to install a decorative border, which would allow the flooring to be aligned perfectly with both the fireplace and the hallway.

Installation near a room's focal point emphasizes the floor's workmanship, be it poor (left) or good (right).

FLOORING DIRECTION DEPENDS ON THE ROOM AND THE SUBFLOOR

After the focal point, the second most important consideration is which direction to run the flooring. Wood flooring generally looks best when the boards are parallel to the longest wall in the room. Running the boards parallel with a shorter wall shortens the room visually.

For structural reasons, strip and plank flooring should be installed perpendicular to the joists. Luckily, this direction generally coincides with the flooring running parallel with the longest wall. If you end

---

FLOORING DIRECTION AFFECTS THE LOOK OF A ROOM

A number of aesthetic considerations might dictate the direction of the flooring. It doesn't matter much in a square room, but in most other cases flooring is usually parallel to the longest wall. Running it perpendicular to the long wall makes a room seem shorter.

1. Flooring direction in a square room does not make too much of a difference.
2. Flooring installed at a diagonal makes the room look bigger, but it can be chaotic in a small room.
3. Flooring installed parallel to a shorter wall can make the room look shorter and wider.
4. Flooring installed parallel to the longest wall makes the room look longer and narrower.

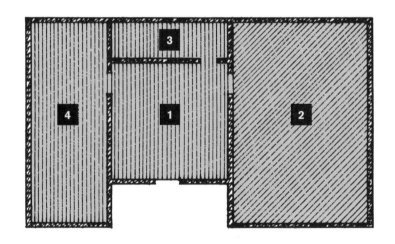

Flooring can be installed diagonally over standard subfloors, but doing so can look chaotic in a small room. Adding a border helps by emphasizing the room's shape.

up running the flooring parallel to the joists, the subfloor must be a minimum 1⅛ in. thick when the joists are 16 in. on center, or there must be blocking between the joists to keep the subfloor from sagging. The blocking is generally spaced no more than 24 in. apart. If the framing is wood trusses, check with the manufacturer for recommendations on how to add blocking.

Strip or plank floor may be laid diagonally to the joists without modifying the subfloor. This can make the floor look larger but also a little busy. I generally do not care for the look of flooring installed on a diagonal unless it is framed with a border or apron.

SEPARATING SPACES SOLVES PROBLEMS

I received one of my most valuable lessons in flooring layout from an elderly man in whose home I was installing flooring. I intended to flow the flooring from room to room, flawlessly tying it back in where I started. When I explained this, my customer's reaction surprised me. Instead of being impressed, he asked, "Why the heck would I want you to do that?" He explained that if I installed the floor that way there would be no separation and he would have to refinish all the floors in the house when any area started to wear.

What my customer realized is that not all rooms in a house will receive the same amount of wear. And even though new wood flooring finishes are very durable, at some time all will require an additional maintenance coat of finish. When the flooring boards run parallel to

the doorway, you can recoat just one room by stopping along the edge
of a board at the door. You can't do this if the flooring runs through the
doorway. For this reason, it is worth considering breaking each room
with a board installed parallel in the doorway.

Separating the flooring layout between rooms can have other benefits.
Often the walls in different rooms of a house don't align. Treating each
room separately minimizes these differences. One example is a long
hallway that directly flows into a main living area. If the flooring runs
continuously between the two, the cumulative effect of even a small
misalignment in the hallway can show up as substantial taper along one
of the other room's walls.

Allowing for Expansion

As explained in chapter 1, solid wood flooring will expand or contract
with any changes in its moisture content. The main issue isn't, as people
tend to think, the floor pushing on the wall and buckling upward. That's
primarily an aesthetic problem and, in fact, upward buckling is a sort of
relief mechanism. The real concern is that if a floor gets wet and
expands, it can exert sufficient pressure to move walls outward and
create a structural problem.

To address this concern, leave clearance between wood flooring and
vertical obstructions such as walls. This clearance allows the last rows

---

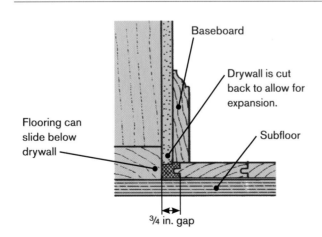

Baseboard

Drywall is cut
back to allow for
expansion.

Flooring can
slide below
drywall

Subfloor

¾ in. gap

CONCEALING FLOORING EXPANSION

Leaving clearance between wood flooring and vertical
obstructions allows the last rows of flooring to expand
with fluctuations in moisture content. Trim traditionally
hides this gap, but with the thinner base molding common today, much of the gap is created by leaving the
drywall higher than the flooring.

of flooring to move enough to pull their fasteners out of the subfloor so that, if the floor grows enough in width to contact the walls, it will buckle upward instead of pushing on the walls. Trim traditionally hides this gap, but with the thinner base molding common today, much of the gap is created by leaving the drywall higher than the flooring.

CLEARANCE FOR SOLID WOOD FLOORING

How much clearance should be left between the wall and flooring generally coincides with the thickness of the flooring material. Typically, ¾ in. of lateral movement is enough to pull the fasteners from the subfloor, so the National Wood Flooring Association (NWFA), the Wood Flooring Manufacturers Association (WFMA), and most manufacturers require a ¾-in. expansion gap left in all vertical obstructions when installing ¾-in.-thick solid wood flooring. This is a safe and simple rule across the grain, though it is a little conservative along the length of the flooring. For example, a 20-ft.-long, ¾-in.-thick solid flooring board is able to change by only about ¼ in. over its length. Someday the expansion gap requirement may be updated to take into consideration that solid wood flooring does not expand the same amount in every direction.

Because fasteners penetrate the tongue of the flooring at an angle, they act like tent stakes and resist a great deal of force in the direction the fastener enters the subfloor. Less force from the opposite direction is necessary to pull the fastener out of the subfloor. For this reason, wood flooring has a tendency to expand mainly in the direction the tongue is facing. In rooms wider than 20 ft., this one-directional movement can cause a problem on the side of the room in which the flooring expands.

Starting the installation in the center of large rooms can reduce these potential problems by forcing the expansion to go in two directions. Since the tongues on either side of the room face different ways, the floor will still expand the same amount, but instead of all the growth focusing on one side, half will go to the opposite side. This reduces the likelihood

Let's Be Clear on Clearance

On a normally fastened wood floor, the clearance around the perimeter of the room has no effect on the boards in the center of the room. If the floor in the center of the room gets wet and expands too much, it will buckle there. The expansion gap at the wall is there to allow the last rows of flooring to release from the subfloor before pushing against the walls. The boards in the center of the room push against each other until the forces are so great that the boards pull the fasteners out of the subfloor and buckle upward.

When the flooring is pushed to the left, the fastener is held firmly in place. When the flooring is pushed to the right, the fastener is easily pulled out. For this reason, areas of flooring tend to grow toward the tongue.

Temporary Spacers Allow for Expansion

When wood flooring is delivered at a lower than desirable moisture content, expansion problems are inevitable. Starting the installation in the center of big rooms can help, as can using spacers. On projects that are expected to have a 4% or more gain in moisture content, install metal washers as temporary shims every three to seven rows. Use washers that are $1/16$ in. thick for strip flooring and $1/8$ in. thick for plank flooring. This technique can also be used to moderate the effects of expansion in floors over 20 ft. wide. Remember to remove temporary washers. For a more accurate estimate of the washer spacing, frequency, and thickness, calculate the expected dimensional change (see Appendix A on pp. 316–323). Proper acclimation is always a more desirable option than the use of temporary washers.

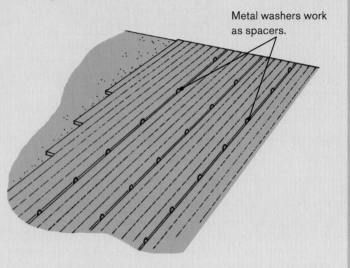

Metal washers work as spacers.

Dealing with Extreme Moisture Swings

A typical home on the coast might have an interior relative humidity of 30% in the winter and 70% in the summer when the windows are open. With that degree of change, each board on a normal strip floor will try to expand by almost $1/16$ in. A floor that is 16 ft. wide has the potential to expand 5 in. The small gaps between the flooring boards will take up some of this expansion, and the fasteners will resist some movement as well.

In part because of the number of joints between the boards, $2\,1/4$-in.-wide strip flooring would perform better than 10-in.-wide planks in that floor. The 10-in.-wide planks would each try to expand $1/4$ in., which might be too much force for the floor to absorb without cupping, buckling, or pushing on the walls. Narrower boards introduce more flexibility to the floor. Engineered, quartersawn, or narrow flooring all perform better than solid, plainsawn wood in environments where there are great fluctuations in humidity levels.

that there will be enough movement to damage the floor. On very large floors, or those that have to be installed at a lower than desired moisture content, it's a good idea to use temporary spacers between the flooring boards. Placing metal washers every few rows (on top of the tongue, so they are easy to remove after nailing the floor) allows for expected expansion (see the top sidebar above).

Installing a hardwood floor is a tool-intensive operation. Fortunately, many of the specialized tools are available to rent.

OTHER TYPES OF FLOORING BEHAVE DIFFERENTLY

Engineered floors behave somewhat differently than solid wood flooring. Although engineered wood flooring changes less dimensionally than solid wood flooring, engineered floors are notorious for developing gaps between the butt ends of boards. This is because most engineered wood flooring expands the same amount in all directions. The lengths of the boards are many times their width, which magnifies any dimensional change along the length. For this reason, it's important to leave the same amount of expansion clearance around the entire perimeter of the floor. Floating wood floors behave even more differently. Since they are unattached to the subfloor, they behave like one giant board and so all the expansion ends up at the perimeter of the room. Consequently, the expansion gap around the entire room is crucial.

Laying Out the Floor

With a simple rectangular floor, the first step in layout, after putting down the vapor retarder (see chapter 3), is to measure the room perpendicular to the direction the flooring will run. It should be the same width on each end.

LEFT AND BELOW RIGHT When walls aren't parallel, the flooring will have to taper. Making up half the taper at each side minimizes the visual impact. Measure the width of the room at both ends. Find the difference between these distances and halve it. Mark the baseline at the narrow end of the room normally. Add half the difference between the widths to the mark on the widest end of the room.

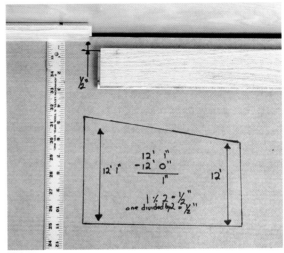

ABOVE Snap a chalkline between the two marks. When the flooring begins on this line, half the taper will occur on each side of the room.

If that's the case, measure out the proper clearance from the wall plate, plus the full width of a flooring board, not including the tongue, and snap a line on the floor. The first row of flooring will align here. (I have a hard time aligning the tongue to a chalkline, so I align it to the bottom of the floorboard instead.)

In many rooms, walls that should be parallel aren't. One side of the room may be 1 in. wider than the other side. If nothing was done to compensate for this, the last row of boards would be tapered an entire inch—nearly half the width of a standard 2¼-in. piece of strip flooring. To minimize the amount of taper required for the last row of boards, it's best to split this difference between the two measurements so it ends up with only a ½-in. taper on each wall. Although a ½-in. taper is still not desirable, it is much better than a 1-in. taper. (Alternatively, you can use a wider board for the first row with its back side tapered to keep the gap small enough to be covered by the trim.)

Installing a Strip Floor

Select the longest and straightest flooring boards for the all-important first row, which should be face-nailed and blind-nailed as per manufacturer's instruction. The old rule of thumb was to face-nail ¾-in.-thick solid wood flooring about 1 in. from the back of the board. New fasteners allow you to get closer without splitting the wood. Face-nail the board every 10 to 12 in. and within 1 to 3 in. from the ends of the flooring boards using 6d or 8d nails. Blind-nail the tongue every 8 to 10 in. and within 1 to 3 in. from the ends using the same spacing.

> Remember, before nailing down the first board, the flooring should be acclimated to the correct moisture content for its environment. Strip flooring should be within 4% of the subfloor moisture content and plank flooring within 2%.

1. Use the longest and straightest boards for the first row.

2. At the end wall, measure off the last board to determine the length of the infill board.

3. After aligning the first row of flooring with the chalkline, face-nail it every 10 to 12 in., about 1 in. out from the wall. Use 6d or 8d finish nails or 2-in., 15-gauge pneumatic nails.

4. A flooring nailer won't fit this close to the wall, so blind-nail the first row's tongue with a 15-gauge finish nailer and 2-in. nails or hammer-driven 6d or 8d finish nails (if using finish nails, predrill to prevent splits).

Face-Nailing vs. Blind-Nailing: What's the Difference?

Blind-nailing means the nails go unseen in the finished job. With flooring, this happens by nailing at an angle through the tongue so the head of the nail ends up well below the floor surface. Face-nailing is just that—you nail through the top face of the flooring and putty the hole before finishing.

RACKING AND NAILING

After installing the first row, it's more efficient to set out all the flooring before continuing to install, rather than selecting a board and nailing it down, selecting a board and nailing it down, and so on. Racking is the process of arranging the flooring boards prior to fastening them to the subflooring. This technique allows you to focus on choosing the right boards, allowing an unbroken rhythm when it's time to start nailing. A 500-sq.-ft. room may have over one thousand boards installed in the floor, and each may vary in length, color, and grain pattern. Arrange the boards in such a way that lengths, color, and grain variations create an aesthetically pleasing pattern. A very light or dark board will stand out like a sore thumb. Working from several containers of flooring at a time helps ensure a uniform color mix. Racking allows you to step back and see the floor from the perspective of the customer. That can help to avoid hearing the question, "How could you install those boards?"

Once most of the wood flooring has been racked out and the first rows installed, the meat and potatoes of the installation can begin. If you're using a pneumatic nailer, keep the airline behind you so it doesn't get in the way. Your shoulder should be in line with the flooring when swinging the flooring mallet, which will keep the mallet landing flat, instead of on an angle.

Generally, you can gently kick the flooring boards into place with the sole of your work boots (soft white rubber soles are a must). If a helper is

FLOORING AROUND AN OBSTACLE

Many flooring installations have to go around obstacles, such as kitchen islands, stairs, or walls, and tie back in together so the rows align perfectly. Snap a chalkline where the two sides tie back in, and measure to it as the installation progresses to ensure the two sides align.

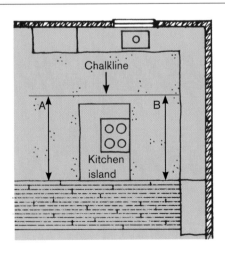

Racking the Floor

Racking is the process of laying out the floorboards in a pleasing manner ahead of installing them. Work from several boxes of flooring at once to ensure a uniform color mix. A word of caution: Manufacturers generally specify that up to 5% of their flooring can have natural or manufacturing defects. Such defects are not covered under warranty. When you begin racking the floor, inspect each board and discard or cut any defects.

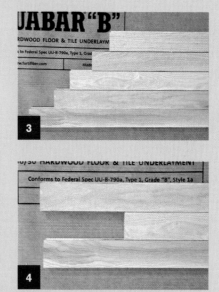

1. Start racking after installing the first row. Leave enough space between the racked boards and the first row to place the flooring nailer.

2. End joints should be spaced at least three times the width of the flooring.

3. Watch out for "stair casing," when end joints in several courses of flooring are spaced evenly so they look like stairs.

4. Another improper racking pattern is when the end joints in three adjacent rows of flooring form an H.

---

available, he or she will already have the boards in place so the nailer will almost sound like an automatic rifle. This is where the installation is the fastest and most efficient.

FITTING THE LAST ROWS

The last rows of flooring are a good indicator of the craftsmanship of the installer. A quality installation will have tight joints, boards that are not noticeably tapered, and show little top nailing. You may need to cut the last board to width, possibly with a slight taper. Although many flooring

(continued on p. 83)

1. Many flooring professionals are able to place boards into position with a careful kick from their soft, white-soled work boots. A careful tap with the rubber end of the flooring mallet fully engages the tongue in the groove.

2. Sometimes a little extra force is necessary to close up the gap between the boards before nailing. Use a small waste piece of flooring or custom plastic block to tap the board into place.

3. In order to save the tongue and groove on the ends of the flooring boards, temporarily turn the last piece of strip flooring in a row around so the scrap end faces the center of the room. The cutoff piece should be less than a couple inches long, or over 1 ft., so it can be used elsewhere.

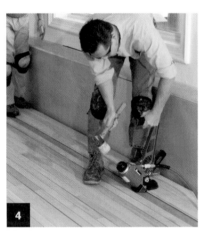

4. You should be able to use the flooring nailer starting with the second or third row of flooring.

5. To get a tight butt joint with the last board in each row, work the board in at a slight angle or use a pry bar at its wall end.

6. Misaligning the flooring nailer will drive the fastener into the top of the flooring. The best solution is to replace the board.

7. To remove a board without damage to the next row in, tap a flat bar under it. Carefully pry up while pulling the flooring slightly toward you.

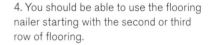

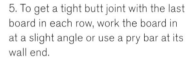

Fitting the Final Rows

1. As you approach the wall, the flooring nailer will no longer fit.

2. Use the straightest boards available for the last rows. The flooring may need to be pried into place. To avoid damage to the wall, place a board behind the pry bar.

3. The last rows are blind-nailed by hand. Predrilling a hole through the tongue of the flooring will keep the nail from bending and the tongue from fracturing. (It is easier to predrill the boards before installing.)

4. Use a nail set to protect the flooring from being hit by the hammerhead.

5. When you can no longer blind-nail by hand, face-nail the remaining boards.

GENERAL FASTENER GUIDELINES

WOOD FLOORING TYPE	FASTENER	FASTENER SPACING
Solid strip • T&G, ¾ in. thick • 3 in. wide or less	• 1½ to 2-in., 16-gauge cleat • 2-in., 15-gauge finish nail • 1½ to 2-in., 15-gauge staple, ½-in. crown • 2-in., 6d to 8d fastener	• Blind fastener through tongue, 1 in. to 3 in. from ends and every 8 in. to 10 in., with a minimum two fasteners per board • 15 gauge finish nails: every 4 in. to 5 in.
Solid strip • T&G, ½ in. thick • 1½ in. or 2 in. wide	• 1½-in., 18-gauge fastener	• Blind fastener through tongue, 1 in. to 3 in. from ends and every 8 in. to 10 in., with a minimum two fasteners per board
Solid strip • T&G, ⅜ in. thick • 1½ in. or 2 in. wide	• 1¼-in., 18-gauge fastener	• Blind fastener through tongue, 1 in. to 3 in. from ends and every 8 in., with a minimum two fasteners per board
Solid strip • T&G, ⁵⁄₁₆ in. thick • 1½ in. or 2 in. wide	• 1 to 1½-in. staples with ⅜-in. or less head • 1 to 1½-in. flooring cleat	• Blind fastener through tongue, 1 in. to 2 in. from ends and every 3 in. to 6 in., with a minimum two fasteners per board
Solid plank • ¾ in. thick	• 1½ to 2-in., 16-gauge cleat • 1½ to 2-in., 15-gauge finish nail • 1½ to 2-in., 15-gauge staple, ½-in. crown • 6d to 8d fastener • #10 deck screw	• Blind fastener through tongue, 1 in. to 3 in. from ends and every 6 in. to 8 in., with a minimum two fasteners per board • Screw, top nail, or glue the ends of each board.
Engineered wood flooring	As recommended by manufacturer	• Blind fastener through tongue, 1 in. to 2 in. from ends and every 3 in. to 6 in., with a minimum two fasteners per board

Note: Always follow the flooring manufacturer's recommendations.

(continued from p. 79)

professionals are accustomed to free-handing this cut on the tablesaw, this is not a safe method. Taper jigs can be made for the tablesaw but do not work well with small portable contractor's tablesaws without a great deal of modifications. Since many times this cut will be covered by trim, it does not have to be perfect. A jigsaw is a safe way to make a taper cut with limited effort and tools.

THE RIGHT FASTENER FOR THE JOB

Solid strip flooring and plank flooring are normally installed using nails, staples, or screws, though wood flooring can also be glued or floated over the subfloor. The fastener size depends on the wood flooring and sub-floor used. Solid ³⁄₄-in. flooring is normally blind-nailed using 1½-in. to 2-in.-long, 16-gauge flooring cleats (flat, barbed nails), or 1½-in. to 2-in.-long, 15-gauge flooring staples with a ½-in. crown. While staples have higher initial strength, in the long run, cleats and staples hold about the same. At the same time, staples have more of a tendency to fracture the flooring's tongue, which negates any holding power.

If top nailing is unavoidable, say for example at the last few boards before a wall, a 2-in., 15-gauge finish nail is generally used. However, there is some controversy over fastening with a 15-gauge pneumatic finish nailer. Although it is the most common method used, it is still not recognized on NOFMA installation procedures, which recommend top nailing with a 6d or 8d nail. A 6d finish nail has a 0.0915-in. diameter (13-gauge), whereas a 15-gauge pneumatic finish nail has a diameter of 0.0720 in., and the holding power of the pneumatic nail is perceived to be lower. Although it is done, I avoid top nailing with 16-gauge flooring cleats because the head of the cleat makes a large, obtrusive hole in the flooring.

Many craftspeople go a step further. Even the holes created by 15-gauge fasteners are too large for them, so they use beads of construction adhesive combined with 18-gauge pneumatic finish nails to fasten their last rows. When using a trowel-applied vapor retarder, the glue will stick to it. If it's paper or some other sheet-type vapor retarder, the area under the first boards is left out. The area that is missing the vapor retarder is so small that it seems to have a negligible effect.

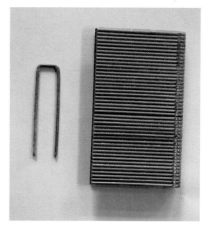

Both traditional flooring cleats (above top) and staples (above bottom) have plenty of grip to hold down wood flooring. The author prefers cleats because they're less likely to split the tongue.

Flooring Nailers: Pneumatic or Manual?

Today most craftspeople install wood flooring with flooring nailers—either manual or pneumatic models. Flooring nailers drive the fastener at the correct angle through the flooring and into the subflooring. The nose of the nailer hooks on the top inside corner of the board's tongue, and striking the pad on top of the nailer with the rubber face of a flooring mallet drives the fastener home.

Some craftspeople prefer manual flooring nailers to pneumatic ones. Manual nailers have fewer parts to break and are not attached to a cumbersome air hose, though they're limited to driving nails, not staples. Pneumatic nailers cost more initially, unless you already own the compressor. But they aren't as tiring to use, installation of the flooring tends to be more even, and models are available to fire either nails or staples. I almost switched to manual nailers several years ago because of the poor quality of the first pneumatic staplers I purchased. Over the years, I ended up buying seven of these staplers in order to keep just a few working. I have since switched to Primatech pneumatic nailers and have had no problems.

Here's another reason I prefer pneumatic nailers. All the force needed to drive the fastener on a manual tool comes from the installer. As the installer gets tired over the course of a day, he or she uses less force resulting in the boards not driven as tightly together. The floor becomes less tight and "grows." And no two installers will fasten the floor with the same force (which is why when two installers are working together, it's a good idea for them to switch positions occasionally to even the floor).

By contrast, pneumatic nailers allow pressurized air to drive a cylinder and perform most of the work, which means it's easier to maintain a uniform floor throughout the day. You do have to make sure to maintain the correct air pressure, generally between 70 psi and 90 psi. Too much air pressure will damage the tongues of the wood flooring when using some nailers.

Whether you opt for manual or pneumatic, either kind is less work than fastening flooring by hand, and either kind is available to rent. Although some old-timers claim they installed more by hand than we do today with all our new electric saws and nailing guns, for most of us a flooring nailer is the way to go.

Whether manual (bottom) or pneumatic (top), both kinds of nailer beat doing it by hand—and either kind is available to rent.

As 15-gauge finish nails have less holding power than 16-gauge flooring cleats or 15-gauge staples, the nailing spacing should be closer, about every 4 to 5 in. In most cases these nails should be long enough to penetrate the subfloor fully, but 1½-in. fasteners are used where they cannot penetrate the bottom side of the subfloor (for example, over radiant heat, or plywood over concrete).

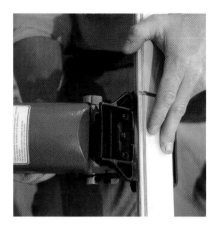
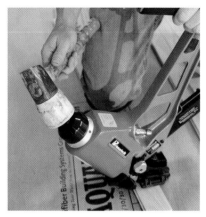

Carefully align the flooring nailer so that it fully seats on the edge of the board (as demonstrated in the photo at far left). Strike the driver on top of the nailer with sufficient force to set the nail fully.

PROPER AIR PRESSURE IN PNEUMATIC NAILERS

The pressure delivered by air compressors is adjustable, usually with the turn of a knob. If the pressure is too low, the fastener won't be driven completely and you'll have to set it with a hammer and nail set. Too high, and the fastener will be driven too deep, resulting in possible tongue damage. Start with the air pressure at 70 to 75 psi and adjust until proper fastener setting occurs. You'll know when the fasteners are driven consistently just below the surface.

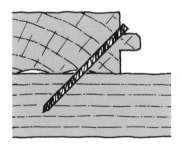

Pressure too low

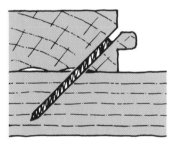

Pressure too high

Correct pressure

Center Layouts

Flooring in most rooms starts from one of the walls, but, as mentioned earlier, it can be a good idea to start the layout of wide rooms from the center so the expansion of the flooring goes in two directions instead of one. To start a center installation, I strike a chalkline down the center of the room, and then screw temporary blocks along it. Start the installation with a row of long, straight boards that butt to these blocks. Just nail through the tongue—there's no need to face-nail this row. Remove the blocks once you've installed four or five rows of flooring.

At this point, only the tongue side of the first row of flooring is fastened to the subfloor. The groove side of the flooring must also be fastened down. The old method of doing this was to face-nail the first row of flooring down the very center of the room, but, to my eyes, these highly visible nails look unattractive. Instead, I install a second tongue, called a spline or a slip tongue, into the groove of the first row of flooring, and nail through that. Although you can't tell by looking at the finished floor, the slip tongue allows the flooring to be installed going in both directions. The spline is about twice the width of the flooring tongue. When installed in the groove, enough protrudes to serve as tongue for the mating board. Spline can be bought from flooring suppliers or made at the job site with a tablesaw.

Glue the spline into place using common PVA wood glue. Do not skip the glue; the spline connection will often fail without it.

In wide rooms, starting the layout in the center and working toward both sides reduces problems with expansion. The starting point is a chalkline centered in the room, which automatically adjusts for out-of-parallel walls.

Screw temporary blocks to the subfloor along the chalkline. The blocks keep the first row straight as it's nailed down.

Once the glue has set, nail through the spline with a flooring nailer, spacing the fasteners in the standard way. The fasteners hold the back end of the board to the subfloor. Splines are used whenever the direction of the flooring needs to be reversed, such as when a floor transitions into a closet or another room (for more on this, see "Making Transitions" on pp. 92–96).

Installing Plank Flooring

Although it is common to install wider boards much like strip flooring, they are more sensitive to changes in moisture content because their greater width allows for more dimensional change per board. Consequently, you need to take some extra measures to ensure a durable installation.

Because plank flooring is much wider than strip flooring, blind-nailing means there are fewer fasteners per sq. ft. holding it down. If 2¼-in.-wide strip flooring is fastened every 8 to 10 in., there will be a little over seven fasteners per sq. ft. Assuming each fastener can resist

After you've installed the first few rows of flooring, remove the starter blocks. Apply wood glue into the groove of the first row.

Insert a spline, or slip tongue, into the glue in the groove and tap it into place.

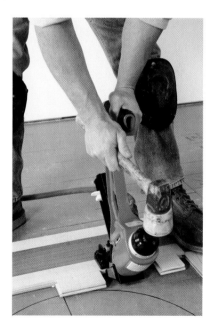

Blind-nail through the spline, using scraps of flooring to keep the spline aligned. Now the flooring installation can proceed in both directions.

Although plank flooring comes tongue and grooved like strip flooring, its width means that blind-nailing alone provides far fewer fasteners per sq. ft. than with strip flooring.

100 lb. of force, each sq. ft. of flooring would be held down with slightly more than 700 lb. of force. The fastener spacing for blind-nailing is decreased on plank flooring to a range of 6 to 8 in., but this still provides about only 400 lb. of force for 7-in.-wide planks, a common dimension. Wider planks would be even less securely fastened.

As a result, plank flooring requires either face-nailing, screws, or adhesive for additional holding power. The additional holding power will help minimize cupping, but it's virtually impossible to stop wood from changing dimensionally with variations in moisture content.

USING ADHESIVE

The easiest way to provide additional holding power for plank flooring is to use construction adhesive. It is easier to apply the adhesive to the subfloor than to the back of the board, so I start by marking the width of the plank on the subfloor. Staying between the previously installed board and this mark, I place beads of adhesive on the subfloor every 10 to 12 in. and 1 in. from the board ends. A ⅜-in.-wide bead of construction adhesive provides about 70 lb. of holding force per linear inch. So a 7-in. bead would hold the plank down with about an additional 490 lb. of force. Added to the roughly 400 lb. per sq. ft. holding power of the blind nails, this brings the holding force per sq. ft. well beyond the 700 lb. of properly nailed strip flooring. Since it is impossible to glue to paper-type moisture barriers, I use Bostik's MVP urethane membrane, which bonds well to construction adhesive.

FASTENING WITH SCREWS AND PLUGS

I like the decorative look of a screwed and plugged plank floor, but each room can require a thousand or more screws and plugs. This requires an incredible amount of labor—drilling and countersinking holes, screw-

Before nailing down the first wide plank board, make sure the flooring has been acclimated within 2% of the subfloor moisture content. Pay special attention to plainsawn planks, as they are more susceptible to cupping than strip flooring or quartersawn planks.

Screwing and Plugging Plank Floors

1. Screws add tremendous holding power, and laying out the holes with a square and a template provides a neat, consistent look.

2. Drill the screw holes, and then countersink them to the correct diameter for the plugs you'll use.

3. When installing the screws, set the clutch on the screw gun to drive the screws snug, but not through the wood.

4. A little white glue holds the plug in place. Pay attention to grain direction and color, depending on whether the plugs should blend in or stand out.

5, 6. Saw off the excess with a thin handsaw, such as a Japanese pull saw.

Adding a bead of construction adhesive every foot or less more than doubles the holding power of blind-nailing a 7-in.-wide plank. (Note that to provide a surface suitable for gluing, the vapor retarder here was applied between the underlayment and the subfloor.)

ing, then gluing in plugs (which I often make myself). But all the work yields the look of an old plank floor held down with wooden pegs. In addition to blind-nailing the tongues, I install #10 deck screws long enough to penetrate the subfloor fully every 32 in. and about 1 in. from the ends. I don't worry about hitting the joists with the screws, though. It takes 221 lb. of force to pull a #10 deck screw out of ¾-in. plywood. Three #10 screws would hold down the plank with 663 lb. of force. The combination of construction adhesive and screws provides more than enough holding power.

Slight variations in the plugs can make big changes in the look of the floor. Plugs cut from dowels show end grain, for a "pegged" floor look, but I prefer to cut plugs from the top surface of a board (face-grain plugs). When installed with the grain in line with the grain of the flooring, the look is subtle. Turn them 90°, and they stand out. For a more dramatic look, cut the plugs from a wood species with a contrasting color. Square holes can be made with a mortising bit and square pegs made on the tablesaw.

For a True Old Look, Use Old Nails

I have restored many wide plank floors dating back to the 1700s. They were all installed with cut nails driven through the top of the boards. There is generally no subfloor, and the nails penetrate directly into the joists.

Cut nails tear through wood fibers, instead of pushing them aside and splitting the wood as modern nails do. The wood fibers push downward and wedge against the nail. Because of this, cut nails have about one and a half times more holding power than modern wire nails of the same length.

Decorative wrought-head nails are still available. I buy mine from the Tremont Nail Company (www.tremontnail.com) in Mansfield, Mass. They simulate the hand-forged nails of the 1700s. The belly nail bows out in the center. As the nail drives into the wood, the broad middle rips a path in the wood but then closes back up around it. This provides even greater holding power.

Cut nails are hand driven and require a little more work to install. When using them, be sure the wedge-shaped face is parallel to the grain. If the wedge shape is driven cross-grain, the board is likely to split. I recommend drilling a pilot hole for the nails placed near the butt ends of the boards.

Modern cut nails with decorative heads are still available.

Colonial-era wide-plank floors were installed directly over the joists, with no subfloor. Cut nails provided effective holding power in some cases for centuries.

TOOLS & MATERIALS

Plug Cutter Extraordinaire

Most plug cutters and counterbores are made from some form of tool steel, but I have not had any luck with them. I have only found one manufacturer I can rely on. Amana Tool® (www.amanatool.com) makes carbide-tipped countersink bits and carbide-tipped plug cutters. The plug cutters have a ½-in. shank and are made for use in a plunge router. These bits are some of my favorite tools in the world and cut through the wood like butter.

A reducer is used to transition between two flooring materials with different heights, typically at a doorway.

You may need to undercut the bottom of the beveled side of the reducer to fit to flooring such as vinyl, ceramic tile, or low pile carpeting.

Making Transitions

Many areas abound in the house where you may need to make a transition between floors of different heights or materials. Typically, these occur at doorways and other openings, but you'll also find transitions within hallways and at the top of stairs. Transitions can be store bought, but because of the variations in flooring thickness and applications, I often custom make them at the job site. Because they are typically installed in highly visible, well-traveled areas of the floor, the material for transitions should be carefully selected. The most common transitions are reducers, T-moldings, and stair nosings.

REDUCERS

Reducers are used to transition between two flooring materials with different heights. They ensure the proper transition when wood flooring meets other floor coverings, such as vinyl, thin ceramic tile, or low-pile carpeting. The top of a transition is flush with the higher floor, and it angles down to meet the other surface. An angle of about 15° is typical, but that can vary depending on circumstances.

T-moldings transition from one hard surface floor to another, and are commonly used in doorways to join two wood floors in adjoining rooms.

1. Make two cuts to establish the thickness of the top of the T, and the width of the leg. If the flooring materials are of different heights, you'll have to change the fence setting between cuts.

2. Make two more cuts to create the T, and rip the upper surface to the desired width and bevel.

3. Use construction adhesive to secure the T-molding to the subfloor. If one edge covers ceramic tile or stone, glue to that also.

4. Predrill the T-molding using a finish nail of the same size you'll use to fasten the molding.

5. Nail through the T-molding and into the subfloor only, allowing the adjoining flooring to move with humidity changes.

Making and Installing a Reducer

1. Set the bevel on a tablesaw to cut the desired transition angle on a piece of flooring stock. Leave the tongue or groove in place as needed to match the new flooring.

2. Sand off the blade marks.

3. If you're installing the transition in a doorway, measure the angle of the jamb to the flooring.

4. Mark the angle on the transition and cut using a miter saw.

5. Predrill and countersink screw holes, bed the transition in construction adhesive, and screw it in place.

6. Use wooden plugs to conceal the screw holes.

Installing Stair Nosing

1. Rip a flooring board to the correct width to finish the flooring and overhang the stair. Glue a strip of matching wood to its bottom edge to match the thickness of the stair treads where the nosing will overhang the top riser.

2. Ease the top and bottom of the nosing with a roundover bit in a router table.

3. You may need to rout in a groove or cut a tongue to join the nosing to the flooring.

4. Use wood glue to bond the nosing permanently to the flooring and construction adhesive to fix the nosing to the subfloor.

5. You don't want to take any chances when installing stair nosing, so reinforce with screws to add holding power and to ensure the construction adhesive bonds to the bottom of the nosing.

With characteristics similar to both T-moldings and transitions, threshold moldings handle larger changes in flooring height. They are ripped to shape on a tablesaw, and then glued and nailed down much like T-moldings.

T-MOLDINGS

T-moldings transition from one hard surface floor to another (see the photos on p. 93). Commonly used in doorways to join two wood floors in adjoining rooms, they can also provide expansion joints when the floor width exceeds 30 ft. Stock for T-molding must be thick enough so its leg can be the same depth as the flooring. The top of the T can't be so high as to be a trip hazard, but it must be thick enough not to break off in normal wear. Usually, ⅜ in. to ½ in. is about right.

STAIR NOSING

Transitioning a floor to the top of a stair is a critical installation. Done wrong, it creates a serious trip hazard. By code, the top of the flooring here is treated as part of the stair. The vertical distance between the treads of a stair must be consistent within ⅜ in., which includes the distance between the upper tread and the flooring. Additionally, the stair nosing must overhang the top riser (the vertical part of a step) by the same amount as the treads overhang their risers. Typically, this is between ¾ in. and 1¼ in.

Glue-Down Floors

Glue provides terrific hold-down power. For example, Bostik's Best moisture-cure urethane adhesive bonds so well it would take 43,200 lb. of force to shear 1 sq. ft. of flooring from the subfloor. But glue isn't the solution for every flooring installation. Most glue-down floors are made from engineered wood flooring, parquet, and some solid wood flooring less than ½ in. thick. These products are generally straight and milled to close tolerances, which is critical if the joints in a glue-down floor are to be tight.

Match the adhesive to whatever flooring you're gluing down. There are three types of wood flooring adhesives readily available on the market today. The most commonly used are water-based pressure sensitive adhesives (PSA) and moisture-cure adhesives. Solvent-based adhesives may also be available, but stricter VOC regulations are making them less common. (For a detailed discussion of adhesives for wood flooring, see Appendix C on pp. 326–328.)

Although modern adhesives are perfectly capable of securing solid ¾-in. strip or plank flooring to a subfloor, it is difficult to find ¾-in. solid wood floor straight enough to be easily drawn up tight while the glue sets. Slightly bowed boards that could easily be made straight with a flooring nailer would require clamping until the glue set, and, of course, there's nothing to attach the clamp to.

Nonetheless, many contractors are developing and refining methods to glue down solid wood flooring. Only the straightest boards are selected for the starter row, and shims are generally used to hold this first row in place so the installer doesn't have to wait for the adhesive to set up. Strip floors are typically installed using a tapping block and mallet. The tapping block protects the flooring but also keeps the mallet from coming into contact with the adhesive. Plank floors may also require ratcheting straps and jacking systems along with plenty of weight to hold down the flooring during curing. No matter what type of wood flooring is being installed, I use 3M Scotch-Blue 2080 tape to hold it together while the glue dries.

INSTALLING A GLUE-DOWN FLOOR

There are two ways to install wood flooring with adhesive. The first method is "wet lay," whereby the flooring is laid directly into wet adhesive. The "dry-lay" method allows the adhesive to "flash" or tack up prior

Match the adhesive to the flooring, in this case moisture-cure urethane adhesive and parquet flooring.

DRY-LAY INSTALLATION

Dry-lay installations have to begin some distance from a wall so the installer isn't standing in tacky adhesive. Snap a line on the floor and spread adhesive beyond that. Once the adhesive tacks up, installation can begin. Once enough of the floor is finished to stand on, spread adhesive over this initial area.

Starting wall

Working area (no adhesive) = multiple of flooring width + expansion gap + 1 tongue width

First row of flooring

Start of next row

Adhesive

ABOVE Setting flooring into the adhesive before it's tacked is called wet lay. It's quicker than the dry-lay method, but you have to be careful not to move the flooring you're standing on.

ABOVE RIGHT Because of dry lay's strong initial tack, wet-ay installations often begin with a dry laid starter strip. Alternatively, temporary blocks screwed to the subfloor keep the first row from drifting.

to installing the wood flooring. The wet-lay method seems to be more popular in today's busy building schedules. As with any flooring installation, a straight and well-secured starting row is key.

With the dry-lay method, installation starts a certain distance (a multiple of the flooring, tongue width, and the required expansion gap) off the starting wall. The area should be wide enough to provide you with just enough working room to start the first rows of flooring, and you'll stand on the newly installed flooring as you go as with a mechanically fastened floor. The initial working area is filled before the end of the installation.

The author keeps wet-lay installations tight until the glue sets with 3M Scotch-Blue 2080 tape. Boxes of flooring add weight.

Wet-lay methods may use temporary shims to align the starting row of flooring to the wall, and the installer works away from the installed flooring, standing on the subfloor. With both wet-lay and dry-lay methods, it's common to install the starting row and then let the adhesive set up prior to continuing the installation. The amount of adhesive applied to the subfloor during a wet-lay installation depends on the length of your arms, the type of adhesive, and how fast the adhesive dries.

On prefinished wood flooring, I hold the boards together with 3M Scotch-Blue 2080 tape while the adhesive sets up (see the photo above). This helps prevent minor shifting or gapping of the flooring. I make sure to clean any adhesive residue off the flooring finish before taping so that it doesn't set up unseen. Be sure the flooring manufacturer approves the solvent you'll need to use to clean up glue squeeze-out. Using improper cleaners or solvents to remove adhesive residue, or failure to remove the adhesive from the floor properly, can damage the finish.

Don't leave tape on the flooring finish longer than 24 hours, or ghost lines may appear after removal. Avoid using regular masking or duct tape, which leaves adhesive residue and may damage the finish. Any tape

Working with Prefinished Flooring

When working with prefinished flooring, it's important to keep in mind that the finish is permanent. While that sounds obvious, flooring contractors used to installing unfinished wood flooring sometimes find it difficult to transition to prefinished flooring. They're used to working directly atop an unfinished floor that will receive a heavy-duty sanding. Protect prefinished flooring from damage. Place tablesaws, miter saws, and hand tools on a clean piece of carpet placed upside down or on something that offers equivalent protection.

Rough handling of prefinished flooring, using improper tools, and improper compressor settings when using pneumatic nailers can damage the edges of the flooring. Avoid striking the edge of prefinished products with the flooring mallet. This can cause edge crushing, unsightly finish cracks, and splinters. Even using a rubber mallet to tap flooring boards with a low-gloss finish into place can burnish the finish. Manufacturers make adapters for flooring nailers that do not damage the boards' edge finish.

Use colored fillers and touch pens from the manufacturer to correct any minor imperfections in prefinished flooring before anyone has a chance to judge the quality of your installation. Do this as soon as possible—once someone sees one imperfection they will search for more.

WATCH THE NOTCH: SIZE MATTERS

Using the correct trowel is critical to ensure the flooring will properly adhere to the subfloor. Trowels are generally designated by the width of the notch ("A"), the depth of the notch ("B"), the spacing between the notches ("C"), and the type of notch. For example, the trowel shown here might be designated a 3/16-in. x 1/4-in. × 1/2-in. V-notch trowel. When using a trowel, hold it at a 45 to 60° angle to the floor, as the angle affects the glue coverage.

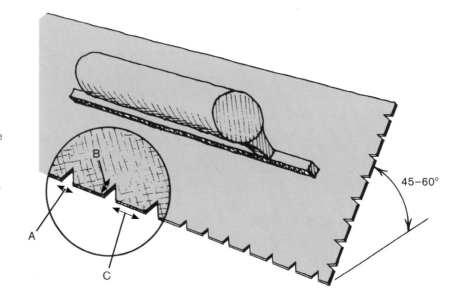

45–60°

left too long on wood flooring finish can cause damage. On unfinished flooring, tape residue is less of a concern, but 3M Scotch-Blue 2080 tape is still the best choice.

THE RIGHT TROWEL FOR THE JOB Although the notch configuration on the trowel you use to spread adhesive might seem like a minor detail, it's not. Many wood flooring failures can be attributed to improper trowel selection. Adhesive manufacturers perform extensive testing to determine the correct trowel to apply their adhesives. The size and profile of the notches control the spread rate of the adhesive, ensuring there's the right amount to hold down the floor. The profile of the notch affects solvent evaporation, which is a factor in the adhesive's open time. V-notch trowels are commonly called for because they promote equal solvent evaporation.

Trowel size is specified to maximize adhesive coverage. Periodically check adhesive coverage during installation. For example, it is recommended when using Bostik's Best adhesive that there should be more than 80% coverage for engineered wood flooring and over 95% coverage for solid wood flooring. This means that if you pick up an engineered flooring board just placed in the adhesive, more than 80% of its underside should be covered with adhesive. Using extra MCU adhesive is never harmful to an installation, as long as it isn't oozing and causing a mess.

Manufacturers generally recommend holding the trowel at a 45 to 60° angle when applying adhesive. A shallower angle would apply less adhesive. A steeper angle might drag part of your hand through the adhesive. And, remember, it's always important to read and follow the directions on the adhesive container.

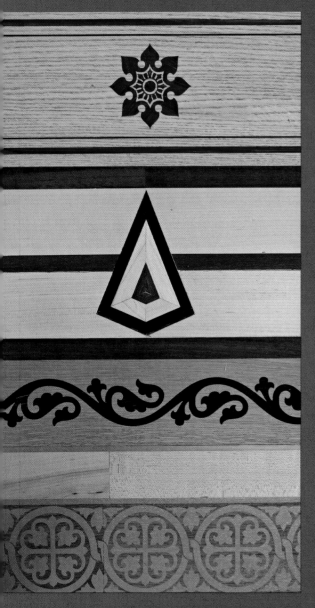

CHAPTER FIVE

An Introduction to Ornate Wood Flooring

THE PREVIOUS CHAPTER COVERED THE MOST common flooring installations, strip and plank flooring. Most of the chapters to come will focus on designing and installing various types of ornamental floors, while this transitional chapter provides an overview of design. The potential designs are virtually unlimited, and the necessary tools and skills are surprisingly few.

As ornate wood floors become more common, designers and homeowners become more aware of options other than strip oak flooring. Computer-controlled machines can economically create medallions and inlays of any design imaginable, and the

Not all ornate floors are as intricate as this one, but using even one of the elements shown here can turn the ordinary into the dramatic.

A variety of decorative elements cut with computer-controlled routers or lasers are available. Manufacturers carry a wide selection of stock design, and most will create any custom design you wish.

schools of the National Wood Flooring Association now teach the skills needed to create ornate wood floors.

A Short History of Ornamental Wood Flooring

No one knows, of course, who first had the idea to lay down an ornamental wood floor, but the designs used to create ornate wood floors are as old as human history itself. From the earliest times, we have decorated our objects, buildings, and even ourselves, finding inspiration in nature, from the shapes of crystals to the symmetry of the veins on a leaf radiating from a central stem.

The traditions of the classical arts that stretched from the Egyptians through the Romans, largely forgotten in Europe, remained in the universities of the Byzantine Empire (see the sidebar on pp. 106–107).

When they took Constantinople in 1453, the Ottomans closed the universities of the Byzantine Empire, and many of their Greek scholars immigrated to Italy. With them came ancient texts and a scholarly tradition that helped fuel the Renaissance.

The Renaissance period in Italy was one of the highpoints of ornamental design, as artisans produced exquisite ornate marble floors in parquetry and marquetry designs. Similar designs are seen throughout the history of wood flooring. Renaissance artisans also produced elaborate wooden panels. It is possible that ornate wood floors were created during this time in Italy, or even earlier, but being subject to wear, fire, and water damage it is hard to find examples more than 500 years old anywhere in the world.

Abstracted and formalized acanthus leaves have been a common ornament for thousands of years. They are traditional in bas-relief carvings and possible to render two-dimensionally in flooring.

THE FRENCH INFLUENCE

Although Italian artisans possessed the skills to produce ornate wood floors, French artisans seemed to lead the way. Ornamental woodwork was generally reserved for areas of higher architectural importance like the ceiling. The ballroom at Fontainebleau Palace was created with just such an elaborate wooden ceiling. In 1547, architect Philibert Delorme went a step further and created a matching ornamental wood floor; it's now considered the oldest remaining significant ornate wood floor in the world (see the bottom left photo on p. 108).

About a century later, the Palace of Versailles, which like Fontainebleau had elaborate wood floors, played a pivotal role in popularizing ornate wood flooring. To consolidate his control, the king of France required nobles to stay at the palace for part of the year, and they began copying Versailles' wood floors at their own estates. Ornate wood floor designs quickly came into vogue and spread throughout France and beyond.

Nobility from other countries were also impressed. England's Queen Henrietta Maria installed parquet floors in Somerset House in about 1650 after her visit to Versailles. In 1717, czar Peter the Great stayed

(continued on p. 109)

The Evolution of Design

The designs used to create ornate wood floors are the cumulative result of contributions from many civilizations. A number of threads that link various civilizations to the common designs we use today can be traced through time.

The illustrations shown here are taken from Owen Jones's classic *The Grammar of Ornament* (London, 1856), which features thousands of drawings of ornamental motifs and design from the ancient world through the Renaissance. Here, we've paired up designs from Jones's book with contemporary versions of various flooring designs. The historic influences are plain to see.

It begins with the Egyptians in about 3100 BC, with designs likely taken directly from patterns in nature. Similarities link Assyrian designs from around 2000 BC to these earlier Egyptian designs.

Greek ornamentation seems to be a refinement of Egyptian and Assyrian designs, while Roman design is an evolution from Greek design, but with less restraint in the use of ornamentation. The Romans lavishly embellished their creations to affirm the greatness of their culture to the rest of the world.

In the mid fourth century, Emperor Constantine moved the capital from Rome east to Byzantium, and renamed it Constantinople. A crossroads of Far East and Mediterranean cultures, Byzantine style reflects the city's multicultural roots.

Egyptian

Assyrian

Greek

Byzantine

Roman

Ornate design can be classified as parquetry, which is geometrical and angular, or marquetry, which uses curved and natural shapes. Both are used in this floor.

TOP RIGHT & BOTTOM RIGHT Inspired by Versailles and Fontainebleau in France, ornate floors spread throughout the Old World in the 18th century. These examples are from Russian palaces of that time.

The floors at Fontainebleau Palace in France date from 1547 and are one of the oldest surviving examples of wood ornate wood flooring.

(continued from p. 79)

at Versailles, which became an inspiration for his own Peterhof Palace. That and other Russian palaces built in the same period house some of the most beautiful ornate wood floors in the world.

FLOORING IN AMERICA

In the United States, ornate wood floors were uncommon in colonial times. Thomas Jefferson (who had been ambassador to France before his presidency) installed parquet floors at Monticello. Despite Jefferson's influence, most homes built prior to 1870 had softwood plank floors of varying widths. Sometimes the floors were painted to make them easier to clean. During the Victorian era, the mass production of parquet wood flooring began. It was commonly installed as a border because oriental carpet covered the center of the room.

American architecture, and floors, trended through a number of fashions after the Victorian era. Largely, custom ornate floors were limited to the homes of the wealthy, while middle-class floors were generally plain or decorated with simple elements such as borders. It was an age of great change, and the mass production of the industrial age made inroads in Victorian flooring. "Wood carpeting," a thin parquet glued to a heavy canvas back, came in vogue. It was mainly installed by cutting sections from a roll and top nailing with numerous tacks (see the photo on p. 110).

Parquet fell out of fashion in the early 20th century, but picked up again in the 1950s with the introduction of cheap tiles that could be directly glued to concrete with asphalt adhesive. By the 1970s, the price of parquet was competitive with carpet, and contractors installed it as fast as they could. The shortcuts taken by some of them gave parquet a bad name, which eventually combined with changing fashion to help it fall from favor.

One of the oldest surviving examples of American ornate wood floors is at Monticello, where Thomas Jefferson designed parquet with a center square of cherry framed in beech. The grain direction of the center squares alternate, and the floors were finished with beeswax.

Parquet Floors

A parquet floor is a nonlinear patterned wood floor. In other words, not all the boards go in the same direction and the pattern is composed of many boards. Parquet floors can be made of parquet tiles or single boards arranged one at a time in a pattern. Herringbone is one of many examples of a parquet pattern (for more, see chapter 7).

Herringbone is a parquet flooring made from short pieces of strip flooring.

A product of the Victorian era, wood carpeting is parquet supplied in canvas-backed rolls and tacked down.

Ornamental Wood Flooring Design

Most ornate designs contain several species of wood, and the differences in color and texture help to make the design stand out. However, mixing different species can be challenging because each species has its own characteristics. It's important to combine only those wood species whose characteristics are similar. Softer woods may show more wear than the harder species surrounding them. Each species reacts differently to moisture changes, which can cause overwood (a noticeable difference in height between abutting boards), gaps, or cupping. Some species may discolor surrounding boards of different species during sanding and finishing. For example, placing bloodwood, a bright red South American wood, next to maple can discolor the maple when the bloodwood is sanded. As long as you pay close attention to details, there is no need to worry. As the saying goes, measure twice and cut once.

In chapter 4, we explained how the layout of wood flooring must take the focal points of the room into consideration, but you also have to consider the placement of furniture. It goes without saying that a decorative wood inlay is wasted if it's hidden under a sofa or chair. At the same time, furniture and flooring borders can be used to define an area such as a dining room without having to enclose it with walls.

SCALE AND PROPORTION, HARMONY AND RHYTHM

The design of the floor should take into account the appropriate scale and proportions of the elements of the room while at the same time maintaining harmony between them. For example, a medallion 7 ft. in diameter might overwhelm a 12-ft.-wide dining room (see the drawing on p. 112).

Harmony isn't easy to describe. The effect is a subjective, I-know-it-when-I-see-it one. One way to describe it is to say that the elements work together. Harmony is achieved through the blend of wood tones and flooring patterns. The patterns can be subtle or bold, as long as there's a balance between the elements.

Rhythm is important to harmony and is created by the recurrence of color, size, or shape in the elements of the design. The eye is able to relax when it recognizes a pattern, even when the overall look is complex. Slight variations maintain visual interest, while using contrasts in color, size, or pattern to bring the eye toward focal points.

Ornate floors don't exist by themselves—they also play a role in the room as a whole. The level of detail in the floor should resonate with that of the room: You wouldn't install a palatial floor in most residences, for

BOTTOM LEFT Combining several species of wood in one floor helps to accentuate the pattern.

BOTTOM RIGHT The most ornate sections don't usually continue all the way to the wall. Borders and aprons around the edges of a room, where furniture is likely to cover the work, are typically simpler than the flooring in the field.

example. One way for the flooring to complement the room is to use its pattern to pull the eye toward focal points such as medallions and borders.

Most elaborate designs help to maintain visual interest and guide the eye through various levels of focal points in the room. Creating a hierarchy among focal points prevents confusion, while employing the subtle use of dissimilar elements helps to maintain visual stimulation.

ELEMENTS OF ORNAMENTAL FLOORING

Most ornate wood floors contain certain basic elements, though by no means does a floor need all of them. The main area of the floor is the field, which can be composed of parquet designs or straight flooring. A single board called an accent strip or feature strip often surrounds the field. This strip is typically made from a wood species whose color contrasts with the field. Its width is generally narrow relative to the proportion of the floor. Borders are decorative flooring bands that define an area such as the field of the floor. They may have a pattern that continues around the room uninterrupted or divided into segments.

PROPORTION IS KING

Even the most beautifully wrought ornate work will look wrong if it's out of scale with the room. The medallion in the drawing at near right is proportional to the field of the floor. The one in the drawing at far right may be dramatic but it dwarfs the space. It's hard to have a sense of what will work in a particular room without drawing the layout on the floor first. In the end, it is the client's preferences that matters.

Corner blocks divide border segments where they meet at corners. Not only do they spice up the border, but they're also helpful to the installer. Without the blocks, whatever pattern the border takes would have to meet up perfectly at the corners. For this to happen, the field would have to be sized perfectly in both dimensions to a multiple of the pattern's repeat. And the pattern would need to be installed with no variations. In the real world, that's very hard to accomplish.

If you eschew corner blocks and want to show off your skills by having the border go around the corners continuously, V-blocks, or center blocks, are a more subtle way to make the joints work out. V-blocks are decorative elements placed midway in the border's length. The border is first joined at the corner, and then installed back toward the middle. The size of the blocks can be varied to work with however the border pattern falls.

Aprons are an outside band of flooring that tie in the entire ornate design to the walls of the room. Aprons are usually relatively plain, often just strip flooring in keeping with their function of picture framing an ornate field. To make it easier to sand the floors, it's a good idea to make aprons at least wide enough so that a big floor sander fits on the apron without overlapping the ornate work in the border and field (see chapter 10).

TOP LEFT Subtle differences in grain orientation and the use of contrasting but similarly warm wood tones are one way to create a harmonious floor.

TOP RIGHT Rhythm is a key to ornate floor design. A floor consisting of a blend of circles and squares and triangles sounds confusing, but the rhythmic placement of such disparate elements results in a cohesive whole.

BOTTOM Bold designs can also be harmonious. Repeated shapes and colors create a cohesive whole out of confident patterns. The dark bands surrounding the field and border set limits that eliminate any visual confusion between the patterns.

1. A stunningly ornate floor complements a Russian palace, and its linear pattern leads the eye to an appropriately heroic-size painting.

2. Design hierarchy keeps a room in scale by telling the eye where to look. The long lines of this floor draw the eye to the distance. Parquetry designs framed in squares provide visual interest and give the eye a few resting places that keep the size of the room from seeming overwhelming.

3. Parquet flooring helps to formalize a grand room in a Washington, DC, federal-style home. The diagonal layout is a visual pull to several focal points such as doors and wall niches.

4. A more modest home calls for a more modest floor. Simple embellishments such as border strips and a center medallion set off this Victorian room beautifully.

Many elements of an ornate wood floor are also found on oriental carpets. The *field* (**A**) is the main area of the floor and can be composed of parquet designs or straight flooring. A *medallion* (**B**) is sometimes inserted as a focal point of the room. Surrounding the field may be a *border* (**C**) with *accent strips* (**D**), *corner blocks* (**E**), *V-blocks* (**F**), and an *apron* (**G**).

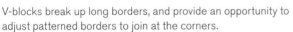
V-blocks break up long borders, and provide an opportunity to adjust patterned borders to join at the corners.

TOP RIGHT Aprons are typically plain and frame the border and field.

ABOVE Corner blocks add another element, but they also save the installer from having to create a border whose pattern has to match perfectly at the corners.

BOTTOM LEFT Accent strips are often of a contrasting wood and outline the field of a floor.

BOTTOM RIGHT Borders provide a visually appealing break between the field and the apron.

Ornate wood floors have a rich history and nearly limitless design possibilities. The following chapters provide additional detailed design and construction techniques, based on the vocabulary from this chapter. The design guidelines discussed throughout this book are just that, though—guidelines—meant only to provide a starting point for your initial design considerations. You can incorporate existing designs into your floor, you can interpret existing designs, or you can create something entirely new.

CHAPTER SIX

Borders & Aprons

IT MAY SEEM COUNTERINTUITIVE, BUT AN ornate floor that runs all the way to the walls of a room can often look less impressive than one surrounded by a simple border. When you walk into a room with an ornate floor like this, no hierarchy exists to tell the eye where to look and the floor can start to seem almost ordinary. Surrounding the field of an ornate floor with a border and apron grabs your attention front and center. It says, "Look here," emphasizing the importance of the field.

Borders and aprons both surround the field of an ornate floor, but they are distinct from each other. A border is a narrow, decorative band that directly

Borders and aprons surround the field of a floor, emphasizing the field and transitioning the floor to the walls.

surrounds the field. Often ornamented with inlays of repetitive forms to create a pattern, a border also functions as a sort of picture frame to set off the field.

An apron is the area between the wall and a border; typically, separated from the border by a feature strip of a contrasting wood. Aprons tend to be less ornate than the border or the field. There are two main reasons for this: First, their comparative plainness accents the ornamentation of the rest of the floor; and second, it's likely there will be furniture around the perimeter of the room that would conceal an ornate apron.

Border Design

There needs to be a balance between the border and the field. You can achieve this balance in many ways, but essentially the border and field should both *contrast* to differentiate the areas and *complement* each other so they don't compete for attention. Borders normally incorporate a species of wood that contrasts with the field. Harmony is achieved

through the balance of wood tones and flooring patterns. Sometimes when the field consists of multiple species of wood, the border will incorporate the same ones but in a different pattern to provide harmony and contrast at the same time. Pattern is a big factor, of course. If the field has a complex pattern, it's common for the border to be simple, and vice versa. This prevents the two areas from visually competing with each other.

When designing a border, you have to keep in mind how you will handle the corners. The range of patterns you can use for a border is practically limitless. However, know that the simpler the border, the simpler it is to install because the corners are easy to join. More on this later in the chapter.

THE GOLDEN RATIO

In design, proportion is king. Make a detail too big and it looks clunky, too small and it gets lost. Good design is largely about how the sizes of the elements relate to each other. Designers search for perfect proportions, but what looks good varies with personal taste. One attempt to quantify aesthetically pleasing proportions is the golden ratio, or golden section, of about 1.6 to 1, which has helped guide artists and architects for thousands of years. According to the golden ratio, the length of a perfect rectangle would be 1.6 times its width—so 5 ft. by 8 ft., or any multiple of those numbers, would yield a perfect rectangle. Numerous studies show that people clearly prefer rectangles proportioned by the golden ratio. However, the golden ratio is of limited use with flooring,

Borders should harmonize with the field without competing. One way to do this is to make most of the border from a contrasting wood, while repeating some of the elements of the field within the border.

Border Patterns

Border designs are often based on ancient design traditions (see chapter 5). They can incorporate botanical, animal, and religious images that have been abstracted and formalized into a pattern—or the pattern may simply be something that appealed to the designer's eye.

The Greek key is a common pattern of the type sometimes referred to as a meander (after Turkey's Meander River).

Abstract botanical designs formalize into a pattern, combining with snowflakes, hearts, and crosses in a visually interesting border.

Accent strips surround a simple geometric design that picks up on the diagonally laid field.

In this Russian floor design, the ratio of the border to the field is about 1 to 6, resulting in a harmonious look and a clear distinction between the two areas.

In another Russian design, the ratio approaches 1 to 2, such that the border competes with the field to leave the eye unsure where to go.

Accent Strips

Accent strips are often incorporated into aprons and borders and are usually prepared from a contrasting wood. They serve to separate the border from the field, or the apron from the border, or just to add a decorative touch in the middle of a border. Combine multiple bands to create simple borders or to enclose complex ones. On their own, they can look great as a single decorative element in a simple floor.

A single accent strip woven into a knot pattern.

A pair of accent strips decorates a simple floor.

because the size of most floors is set long before the flooring is installed. You can put it to use occasionally in ornate floors that contain multiple rectangles and in some border designs.

The width of the border and apron should be in proportion to the size of the room and the flooring material used for the field. Although there are no fixed rules as to the proportions of the border to the field, the size and color of the border should be in balance with other elements in the floor. My eye is partial to the classic proportion used for interior trim, which is 1 to 6. By this ratio, a door with an opening of 36 in. would have a trim width of 6 in., half on each side of the door. This would make the entire width of the window and trim 42 in. The door is akin to the field of a floor, and the trim to the combined border and apron. Remember that there are seven parts in a 1 to 6 ratio. The door is six units and the trim one unit, for seven total units. Figuring the field and border/apron of floors differs from figuring doors and trim. The total width of the floor is fixed, and to find the width of the parts of the floor you first divide the total width by 7. The field will measure $^{6}/_{7}$ of the total width, and the width of the border and apron will be half of the remaining $^{1}/_{7}$. For example, consider a 21-ft.-wide room, dividing 21 ft. by 7 yields 3 ft. The field would be 18 ft. wide (6 x 3 ft.) and the border/apron would be 1 ft. 6 in. wide (3 ft. divided by 2).

Of course, what looks good to me may not look good to you. You should consider the 1 to 6 ratio to be a starting point rather than an inflexible rule. That said, I don't think I'm alone in appreciating this proportion—it's often found in Oriental rugs where the range seems to be from 1 to 4 or 1 to 6.

Aprons

Aprons are one of my favorite elements on a wood floor. In most cases, aprons are more utilitarian than borders. They both provide an area to place furniture without covering up the ornate work in the field or border, and at the same time focus attention on those other areas. From a design perspective, aprons can provide an area in which the eye doesn't expect much. This is important because it provides a workable area that can compensate for discrepancies in the walls or the room's proportions. For example, if the wall jogs slightly, the apron can accommodate this irregularity and the border and field can remain a rectangle.

In those cases where the walls jog in and out or are severely out of square, using a more complicated flooring pattern for the apron can work well. The narrow slats typically used for herringbone floors can be laid perpendicular to the wall or installed as you'd expect, in a herringbone pattern. You can use standard strip flooring this way, as well. The advantage of any of these installations is that they hide flaws in the wall.

Corners where the apron meets can be mitered or laid log-cabin style. Variations on these themes include reverse log-cabin corners and chevron corners. Add decorative elements to the corners for more interest.

Making and Installing Borders

Many borders with angular patterns can be made with a simple crosscut sled and a tablesaw (for directions on making the crosscut sled, see pp. 144–145). The sled makes precisely repeated cuts that are accurate to thousands of an inch to the small pieces of wood that are often necessary for a border. It does all this and keeps your fingers safely away from the saw blade. Borders can be made from either full-thickness flooring material, or from $5/16$-in. veneers glued to a $1/2$-in. plywood substrate

(continued on p. 125)

LEFT Aprons provide an area to make adjustments so the floor can flow smoothly to the wall.

ABOVE LEFT Laying strip flooring perpendicular makes it difficult to see unevenness in the wall.

ABOVE MIDDLE When aprons run perpendicular to the wall, one result is the reverse log cabin corner.

ABOVE RIGHT Patterned aprons help transition to the wall.

BOTTOM LEFT Log cabin corners where the aprons meet are traditional.

BOTTOM RIGHT Running the apron on a diagonal allows for a chevron corner.

Installing a Herringbone Apron

Using a herringbone pattern apron is a good way to hide the fact that a room is not perfectly square. You can purchase precut herringbone flooring for this purpose, but the available lengths may not work for your situation (see chapter 7 for more on herringbone floors). In the case shown here, I laid a simple chevron design, deciding the length of the slats based on what was necessary to fill the apron and cutting the pieces on a miter saw. Cutting the slats a consistent length and angle was still important, so I set up stops on the saw to ensure precision.

I began the layout in the middle, and ran the chevrons in opposite directions to the corners. I prefer to glue down herringbone, but, with it being as close to the wall as it was, glue was practically a necessity. There simply wasn't much room to swing a hammer. To help ensure a tight, long-lasting joint with the border, I splined the ends of the herringbone.

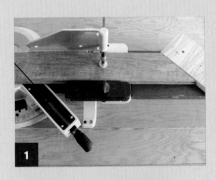

1. Miter the slats using an edge guide to ensure repeated accuracy.

2. Use a stop block and a clamp for precision. Clear away any sawdust from the fence and stop between cuts.

3. Start the installation against a plywood right triangle screwed to the subfloor and squared to the border.

4. Glue the chevron down with urethane adhesive.

5. Drive a few nails in each slat to keep it from moving while the glue sets. This won't be easy due to the proximity to the wall.

6. Spline the joint for extra security.

7. Work the border from the center toward the end. The direction of the slats will allow the perpendicular side to meet this side in a 45° corner.

(continued from p. 122)

(for more on this, see chapter 7). Make up a prototype of the border pattern first, then cut all the pieces you'll need plus a few extras to allow for culling bad stock. When cutting, always be sure of the placement of your hands, and turn off the saw before reaching to clear the cut pieces.

WORKING WITH REPEATS

If you've ever hung patterned wallpaper, you'll be familiar with the concept of the repeat, which is the distance a border goes before the pattern repeats itself. For the pattern to repeat so that it meets in a perfect miter at the corner, each side of the field plus two border widths must be a multiple of the repeat. In other words, side A plus two border widths and side B plus two border widths must be evenly divisible by the repeat.

This hardly ever works out without some adjustment, and the math could drive a person crazy. Instead, I typically lay out the border first and mark on the subfloor approximately where it will go. Then I lay the field using the borderlines as a guide. Once I finish installing the field, I lay the border pieces in place so I can determine the exact final size of the field. Borders with shorter repeats are far easier to adjust, because the adjustment is never more than a fraction of the repeat. Once done, I trim the field to size using a track saw (see the bottom photo on p. 131). If the border has a tongue, I rout a groove in the edge of the field after trimming it. The border should now fit perfectly in place.

You can add feature strips between the field and the border to adjust where the repeat joins at corners. Or, you can simply use corner blocks to avoid the need for a mitered corner. When I make the border, some patterns lend themselves to adjustment in the middle of the run. Alternatively, you can use blocks in the center of the border to adjust its length so the corner miter works out.

INSTALLING A CENTER BLOCK

Even simple borders with no pattern can benefit from a center block. In the photo sequence shown on p. 126, I'm installing a border made from purchased feature strips of bloodwood and Peruvian walnut and two strips of quartersawn maple. The center block also serves to hide a joint in the feature strip. After squaring the end of the border, I lay it in place

1. Make simple 45° cuts with a fence screwed to the crosscut sled.

2. Use stops to ensure precise lengths for repeated cuts. Make all the cuts for each particular length at one time.

3. A crosscut sled on a tablesaw cuts small pieces safely. Attempting similar cuts using a miter saw would send most of the pieces flying into the dust on the floor.

4. Use a second fence and stop for rectangular pieces.

and trace the shape of the center block on it. Disassembling the border, I cut the individual strips to the correct angle and assemble the border and center block.

MITERING THE CORNER

That brings me to mitering the corner, which is much better done in one cut with the pieces of the border all assembled. Because miters tend to open up with seasonal movement (and the wider the board, the more this becomes noticeable), I spline the corner. I lay the border in Bostik's Best, and hold it in place with blocks screwed to the floor until the glue cures.

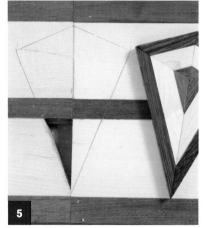

1. Mark the center of the border for cutting.

2. Assemble the border pieces and cut them all at one time to ensure alignment.

3. Hide the joint with a V-block. Dry-lay the border and mark the block's location.

4. Disassemble the border and cut the pieces individually.

5. It is best to make the two-sided cut on the upper piece of maple on a tablesaw crosscut sled.

TOP LEFT After dry-laying and marking the border in place, miter it and set it in place. Screw temporary blocks to the floor to hold the border while routing for a spline.

TOP RIGHT Spline the joint using carpenter's glue.

BOTTOM Glue the border to the floor and secure it with temporary blocks.

CHAPTER SEVEN

Parquet Floors

WHEN PEOPLE HEAR THE WORD *PARQUET*, THEY often think of the cheap and simple wood tiles installed by the acre in the 1960s and '70. But there's more to parquet floors than that. Broadly speaking, parquet floors are a mosaic of wood pieces combined to form a pattern. They have graced palaces and historic buildings for at least 500 years and are considered by many people to be the most beautiful wood floors ever created. Thousands of patterns exist, and the range of possibilities is limited only by your imagination.

Parquet floors can be made either by installing one board at a time, as is typical with herringbone floors,

TOP LEFT Parquet floors combine a variety of wood species and shapes into decorative patterns.

TOP RIGHT Herringbone parquet flooring is normally installed one board at a time.

BOTTOM LEFT Inserting a dark block of wood into this design creates negative space for a basketweave effect that fools the eye.

BOTTOM RIGHT Some classic parquet patterns combine smaller pieces of flooring into large mosaics, as in this floor from Versailles.

Parquetry, Marquetry: What's the Difference?

Traditional parquetry patterns are geometrical and angular. Add curves and other natural shapes, and the patterns become marquetry. Parquet designs are easy to make with a tablesaw, but marquetry requires great skill cutting curved elements with a scroll saw. Both parquetry and marquetry designs are available as manufactured wood tiles.

or by installing wooden tiles composed of smaller pieces. Some of the oldest parquet tiles are held together with painstakingly cut mortise and tenon joints. With the high-quality adhesives available today, many of the parquet tiles I make are simply held together with tape or temporarily adhered to a paper backing until they're bedded in glue. (Because I glue down parquet, the vapor retarder is always a trowel-applied product that's compatible with the adhesive.) Some parquet flooring is made from standard ¾-in.-thick flooring, while other parquet consists of thinner material. In this chapter, I'll cover making and installing several types of parquet, beginning with ¾-in. herringbone and including parquet tile and a three-dimensional-looking rhomboid floor.

Installing parquet flooring requires more layout work than strip flooring. Sometimes, it just entails snapping more layout lines on the floor, but you may also have to screw some plywood straightedges in place temporarily to provide backing. It depends on the type of flooring you're laying. If there's a border, I usually install the parquet in the field first. I try to manipulate the size of my feature strips, border, and apron so I do not have to trim the parquet. In some cases, the parquet may have to be cut. I draw working lines on the floor prior to starting the installation so I know where *not* to apply the adhesive. After installation, I measure and redraw cutting lines onto the field of parquet before installing the border.

Instead of installing the entire border first and having to fit every parquet tile to it, it's better to install the parquet tiles so they lap the joining line. A track saw trims the parquet tiles to the exact size to meet the last board of the border.

Herringbone Parquet

Herringbone, the classic parquet design, consists of flooring laid in a zigzag or V-shaped pattern. Herringbone parquet generally looks best when you install the points of the Vs parallel to the longest direction of the room. As with any floor, consider focal points when deciding the layout (the points of the herringbone should align or be symmetrical with the focal points). To make sure my customers understand how the floor will look, I always dry-lay a few rows before starting the permanent installation.

The potential variation of herringbone designs is huge. The tips of the boards, or slats, can be mitered or square-edged. Mitered herringbone slats are sometimes referred to as chevrons or Hungarian-style herringbone. The slats themselves can be made of multiple bands of different color woods. One of my favorite looks is to square-cut the ends of the slats and add a square of a contrasting wood at the point of the V instead of overlapping the ends. Because of the angular nature of herringbone floors, this square block looks like a diamond. In the floor shown in the installation sequence on p. 136, the diamond is white oak and the slats are Brazilian cherry.

Although it doesn't take a great deal of skill to install a herringbone floor, it does require a lot of attention to detail. The subfloor must be flat within ⅛ in. over 6 ft. or the pattern will change with the highs and lows in the subfloor. The herringbone slats must be dimensionally exact or the pattern will wander off course. Because of this, snapping or drawing layout lines on the floor is a critical step.

LAYING OUT HERRINGBONE FLOORS

When laying out a herringbone floor, there are a few things to bear in mind. First, when ordering material figure on about 15% waste, which is about three times as much as I'd allow for a standard strip or plank floor. Next, you'll need to decide the desired width of the floor's apron, if any (see chapter 6), and from that determine the rough width of the field where the herringbone will go. You can run a herringbone floor all the way to the walls, but often enough furniture will cover the floor within 3 ft. of the walls, which may conceal all your hard work. Also, surrounding ornate floors with a border and apron sets them off, much

as a good frame displays a painting to its best advantage. The width of each V-shaped section looks best when it divides evenly into the space between the borders. That's not likely to happen the first time you run the numbers. The length of the slats determines the width of the herringbone: It has to be a multiple of the width of the slats. Complicating all of this is the fact that the slats run at a 45° angle to the width of the room, and you need to drag out the Pythagorean Theorem to figure it out mathematically.

Instead of doing the math, I find it easier to decide the approximate width of the herringbone and lay out some sample slats on the floor that I can fine-tune so they're a multiple of the flooring width. Once I dial in the exact width of the herringbone, I can then adjust the width of the border to fit. Remember too that the final width of the herringbone field will be trimmed in place with a track saw, so it's easy enough to narrow the field by an inch or two. That said, if you don't have a track saw, it's possible, though not as efficient, to make accurate cuts by guiding a circular saw with a plywood straightedge screwed to the floor.

Because it affects the layout and the length of the slats, it's crucial that all the flooring be exactly the same width at installation. To ensure

With its overlapping, angled slats, herringbone is a timeless parquet design. This example is in Washington, DC.

HERRINGBONE LAYOUT

The length of the individual slats that comprise herringbone flooring must be a multiple of their width.

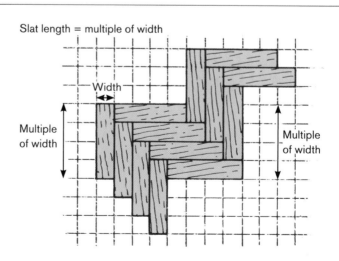

Slat length = multiple of width

Width

Multiple of width

Multiple of width

this, I run all the stock through a tablesaw, cutting off just as little as possible from the groove edge to make all the boards a uniform width. Taking any width from the tongue edge is more complicated—you'd need to make a shallow cut from each side of the board to leave the tongue intact. From the groove edge, you can take off a small amount of wood in one pass with no further action required. Be careful here: Because tongue width may vary, and the tongue is what registers on the fence, you may need to change the saw's setting if you're mixing flooring from different mills. And if you rip a lot from the edge, you may need to re-groove the boards. Traditional installations of herringbone flooring also require that the end of each slat have a tongue or groove to interlock with the edge of the abutting slat. That's a lot of work, and it's rendered unnecessary by laying the flooring in adhesive.

Marching all these angled boards across the room in a straight line would be impossible without accurate layout lines on the subfloor. If mitering the herringbone, I mark a centerline for each angle, starting from the center of the room (see the drawing on the facing page). For square ends or with a diamond inlay, it's a little more complicated. In this case, two lines per angle are necessary, one for each corner of the ends. The space between these two lines is the length of a line drawn at 45° across a slat.

The slats in herringbone flooring must be a multiple of their width. Before production-cutting the stock for a floor, use this quick check to verify the slat length.

Precision is key to successful herringbone installation. Once you're sure of the slat length, you can cut them all with a miter saw and a firmly attached stop block.

Obtain the distance between the layout lines for square-edge herringbone by measuring between two 45° lines that pass through the corners of a pair of herringbone slats.

LAYOUT LINES DEPEND ON THE JOINERY

Installing a herringbone floor requires working lines to keep the pattern straight and regular. With square-edge herringbone, line A marks the center of the subfloor and is necessary only for the first line of joints. All the other lines lay out from this one. Each set of joints requires two lines, B and C, one for each corner of the boards.

With mitered (also called chevron or Hungarian) herringbone, line A orients the first line of miters, and single working lines lay out from it.

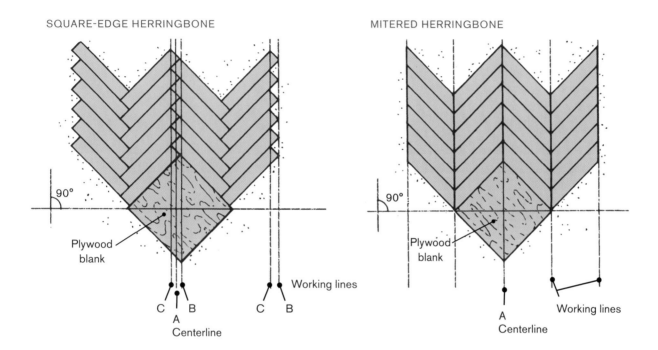

SQUARE-EDGE HERRINGBONE

MITERED HERRINGBONE

90°

90°

Plywood blank

Plywood blank

Working lines

Working lines

C   B   C   B

A
Centerline

A
Centerline

Once I've snapped all the lines on the floor, the next step is to cut the slats to length. Precision is key here again, and using a miter saw with a stop block ensures accuracy (you can also use a crosscut sled, as described on p. 144–145).

INSTALLING HERRINGBONE FLOORING

Starting on layout is critical, but the lines on the subfloor are only a part of the equation. You also need something firm to back up the boards as you nail them. Without firm backing, there's no way to blind-nail the first few courses without having them move out of position. I provide

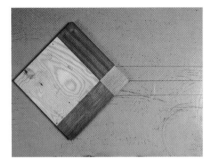

A square piece of plywood set so the centerline bisects it provides both backing and alignment to start a herringbone installation.

1. Begin installation in the center of the field. Spread flooring adhesive. At the edges, don't spread adhesive beyond where you will trim the herringbone.

2. Lay the herringbone tongue side out. Although not strictly necessary, driving one or two nails in each slat keeps them from moving accidentally before the glue sets.

3. Set the diamond between the slats.

4. Continue installing the slats and diamonds on the layout lines, tapping as needed to set them tight.

5. At the end of the row, it helps to screw additional square starting blocks to the floor.

6, 7. Spread just enough adhesive for the next row of herringbone, and begin laying the slats going back the way you came. Install succeeding rows in opposite directions.

this backing with a square piece of ¾-in. plywood that's about the same length as the slats screwed to the floor.

Working one row at a time, spread out the adhesive (I use Bostik's Best moisture-cure adhesive; see Appendix C), and nail the first slats against the plywood block. Be sure to scrape off the excess adhesive as you go, or it will harden. (If that happens, you'll have a more difficult time removing it, which you must do or the next slats won't sit flat on the subfloor.) The next row also begins with a piece of plywood, and I reverse directions working back toward the starting point. Continue in this pattern until the field area is filled. Once the row is complete, I remove the plywood starting block, and fill in as needed with slats. Any slats used to fill in here will be running in the opposite direction, so you'll need to use a piece of slip-tongue or spline to reverse direction. I stop the glue and the nails at the borderline, let the edge pieces run past, and trim them off later with a track saw. Pay attention when nailing these pieces so that you don't place fasteners in the path of the saw. (In most cases, parquet is just glued down, nails are used only occasionally and mainly to hold the initial alignment.)

Once I've trimmed the herringbone all around, I proceed to install the border (see chapter 6). If there's no border, each board will need to be cut at an angle to meet the wall, which can be more labor intensive. At minimum, you'll want to set up a miter saw with a stop block to make these repeated cuts.

Parquet Tiles

Installing parquet tiles creates the most spectacular floors for the least amount of effort. If you take a couple of essential precautions, the floor should come out perfect. First, the subfloor must be flat within ⅛ in. over 6 ft. Subfloors that resemble a roller coaster make it impossible to keep the pattern aligned. Second, primary and secondary lines must be laid out with great accuracy. While these points are true of any wood floor, they're particularly important for parquet flooring. Unlike strip or plank flooring, which is linear, parquet flooring is composed of multiple smaller pieces, and small deviations can easily cause the flooring to wander off the straight path.

Parquet tiles can be made from ¾-in. flooring temporarily held together with tape or glue. When gluing the tiles to the subfloor, remove the tape.

LAYING OUT TILE FLOORS

Laying out a parquet tile floor is somewhat different than laying out a parquet floor that's installed one board at a time (as was the herringbone floor we just discussed). The most efficient method is to start from the center of the room and work toward the walls. If installing the apron and border first, every piece of flooring in the field would have to be meticulously cut to fit inside them. It is much easier to install the field first, trim anything needed with a track saw, and then install the border around it.

Most rooms are not perfectly square, and some of the walls may not be straight. Snap a centerline in the room and check the room's layout from it. The main objective is to keep the parquet squares symmetrical with the walls. Place enough parquet squares to go across the room, and adjust the layout if need be to work with the focal point and at the walls. These adjustments usually aren't big, or much different from a strip floor. For example, moving the layout over an inch might center it on a doorway or a fireplace. Make up the difference by varying the layout on each side of the room. Because of the distance across the room, no one will notice this discrepancy. Make these adjustments, and then find the center of this line. Mark exactly in the middle between the other two walls, then lay out some parquet squares to see how they work in

Parquet with Borders

As explained in chapter 6, a border is typically an ornate band of flooring installed on the outer periphery of the floor (there may also be a plainer apron between the border and the walls). The border serves as a transition for the decorative floor to the walls. The borders (and aprons) can add beauty to the floor and help make the floor installation easier. You can adjust the width of the border or apron to compensate for many design considerations. It can help hide walls that are not straight, reduce the number of parquet tiles that need to be cut, help transition into another room, or incorporate a focal point with the parquet floor.

this direction. If you can't get a layout that pleases you, there is a good possibility that the floor would maybe look better surrounded with a border and apron anyway.

Square parquet requires two main working lines, a center layout line and a second line perpendicular to the primary line. There are a couple different ways to create the second line. You can use a laser that provides lines of light 90° from each other. Lasers are quick and easy, but the thickness of the light line widens with distance. You can also use the 3-4-5 triangle method to create a secondary square line. Any triangle whose legs measure 3 and 4, and whose hypotenuse measures 5, is a right triangle. You can increase these numbers by any consistent multiple to make a larger, more accurate triangle. So, by marking off a 4-ft. section of the centerline, and finding where lines measuring 3 ft. and 5 ft. from its ends intersect, you can construct an accurate right angle. That said, this method is easier to use for verifying square than for laying out a square corner. My favorite method is to use trammel points to swing intersecting arcs, which create square lines 90° from each other (see the sidebar on p. 140).

Trammels turn any length stick into a compass.

INSTALLING TILE FLOORS

Once I'm happy with the layout and snapped the primary and secondary lines, I snap or draw a full grid that locates each parquet tile. Then, I screw some straight lengths of plywood to the primary and secondary lines to provide backing for the first parquet tiles installed. Otherwise, they'd slip around on the adhesive and be difficult to keep on layout.

Starting in one quadrant, apply adhesive with a notched trowel. Many types of adhesives are available for parquet floors. Urethane adhesives hold well over many years of floor expansion and contractions due to moisture changes, and they're what I use in nearly every instance (for more on adhesives for wood floors, see Appendix C on pp. 326–328). Apply the adhesive per the manufacturer's instructions, and start laying parquet tight to the plywood backers.

When you've completed the first quadrant, remove one of the backer boards. The second quadrant will use a backer board on one side and the already installed parquet squares on the other side for alignment. Do not walk on the floor until the adhesive has set (setting time varies

Using Trammel Points

Trammel points turn a stick into a compass and are the most accurate method to lay out secondary lines at 90° off a primary line. If you took geometry in high school, you may remember constructing figures using a compass. The method is identical, though the scale is somewhat larger.

Trammel points consist of clamps that affix to a stick (photo at top right). One clamp also holds a steel point and the other holds a pencil. The point is placed at the center of a desired arc, which is drawn by the pencil.

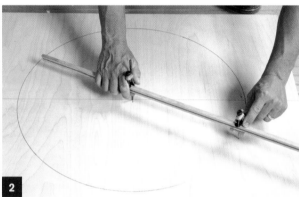

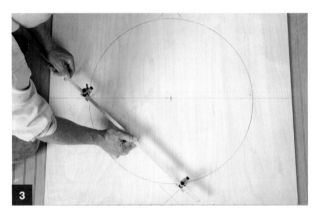

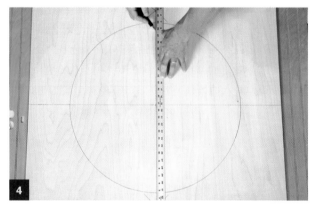

1. Mark the center point of two parallel lines (or walls) and draw a centerline that connects the two marks.

2. Draw a circle that originates at the middle of the centerline. Make it large enough to draw an accurate line between two points opposite each other—say, half the width of the area or more.

3. Extend the distance between the trammels to swing longer arcs between the intersections of the primary line and the first arc.

4. Connecting the intersecting arcs creates a secondary line that is square to the primary line.

1. Parquet squares can wander off course during installation. Establishing a grid on the subfloor with a chalkline or with a pencil and straightedge is a great help. Each square of the grid should correspond to the size of the individual parquet tiles.

2. Temporarily screw straight plywood boards on the primary and secondary lines to serve as backer boards to keep the parquet tiles aligned during initial installation. Install the parquet in quadrants, starting in the center of the floor and working away from the backer boards. Here, the author lays out an installation grid from these backers.

3. Trowel on adhesive to as large an area as you can reach that you'll be able to cover before the adhesive sets.

4. Gently work the parquet in so not to trap the adhesive between the edges of the parquet squares (here, the author installs paperface parquet). Work the parquet gently back and forth to ensure proper adhesion. Use the trowel to scrape off any adhesive that stiffens before you have a chance to install over it.

depending on the adhesive, so check the manufacturer's instructions). Leaving room for a border and apron will allow access around the room while the adhesive is curing. In other words, do not parquet yourself into a corner. If there's no apron, then be sure to work toward an exit.

Three-Dimensional Rhomboid Parquet

By combining wood species of slightly different colors and varying the alignment of the wood grain, you can create dramatic three-dimensional effects. The example shown on the following pages creates a floor that looks like a series of abutting cubes. Each "cube," or "rhomboid," is a parquet tile made from three identical rhomboids whose adjacent corners are 60° and 120°. The rhomboids are of different species (ash, maple, and walnut, in this case), and each is cut and placed so that the

Properly called rhomboids, the cubic patterns in this parquet combine wood of three species to create a three-dimensional effect.

RHOMBOID PATTERN

The individual rhombuses that make up the cubes, or rhomboids, in a floor only fit together if their corners are exactly 120° and 60°.

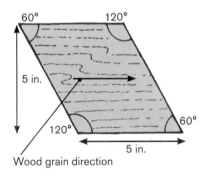

Wood grain direction

grain aligns in thee different directions. This arrangement helps to lead the eye around the individual "cubes," while helping them to stand out from each other.

As ever, a few precautions can save a great deal of inconvenience later on. First, make sure the boards of the three species are the exact same width or the segments will not fit together. It is also important that the three species are at the same moisture content or gaps will occur when they equalize. Use a moisture meter to verify the moisture content (see chapter 2). Last, try to pick species that respond similarly to changes in humidity (see chapter 1). One way to get around this potential problem is to make the parquets from ⁵⁄₁₆-in.-thick veneer (ripping ¾-in. flooring in half nets two pieces of ⁵⁄₁₆-in. veneer) and gluing them to Baltic birch plywood backing.

CUTTING THE PIECES

The first step is to cut the individual pieces for each rhomboid. The rhomboids have to be on the money—only angles of exactly 60° and 120° will work—and the lengths and widths of each piece must be identical. The best way to make the rhomboids is to use a crosscut sled with its fence set at 60° (see the sidebar on pp. 144–145). To ensure their length is consistent, use a stop block. After cutting the first three pieces, I test-fit them. The fence often requires micro-adjustment. When cutting the rhomboids, housekeeping is important. It doesn't take much sawdust between the fence and the stock to throw off the angles.

To cut the first rhomboid, set the fence as close as possible to 120° to the saw blade. The resulting cut will be the reciprocal angle of 120°, or 60°.

Use a cutoff from the rhomboid stock to set a stop block the correct distance from the blade.

Test-cut a few pieces for assembly into a rhomboid.

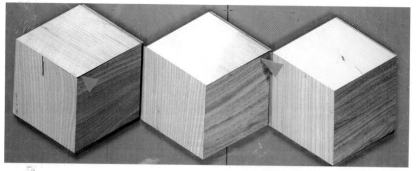

It often takes several tries to cut rhomboids that fit perfectly. The first rhomboid has a gap in the left-hand corner. The angle on the fence was adjusted 0.008 in., which was too much and left a gap in the upper right-hand corner of the second rhomboid. The third rhomboid was adjusted back 0.004 in. and fit together perfectly.

Make sure to clear sawdust from between the fence and cutting stock, otherwise you compromise the accuracy of your cuts.

ASSEMBLING AND INSTALLING THE RHOMBOID TILES

Once the fence is set, cutting consistent rhomboids is relatively mindless work (but don't let your mind wander—you're working near a spinning blade). Assembling the cubes is simplicity itself. I use regular PVA carpenter's glue, and clamp the rhomboids together with tape. I violate all sorts of glue-up rules in the process—gluing end grain, *not* worrying about squeeze-out, and having minimal contact area where a grooved edge meets another rhomboid. Two things prevent these issues from being problems. First, the tiles will be set in adhesive, which will hold them together so well that the edge glue joints don't matter. Second, heavy floor sanders make quick work of glue squeeze-out that would take forever to sand out of a furniture project.

A Crosscut Sled Makes for Consistent Accuracy

A basic crosscut sled turns the cheapest tablesaw into a precision tool capable of making cut after cut that's accurate to thousands of an inch. It does all this and keeps your fingers safely away from the saw blade.

A crosscut sled is nothing but a piece of plywood or MDF that slides across the top of your tablesaw. I prefer to make sleds from quality birch plywood so they will last a long time. I also apply a coat of polyurethane to the sled to help prevent moisture from distorting it over time.

Two runners on its bottom side slide in the miter-gauge slots on the top of the tablesaw. The runners keep the sled perfectly aligned to the saw blade. I generally cut the runners from a piece of leftover oak strip flooring. Stabilizers on either end of the sled hold it together, and the front stabilizer doubles as a fence. You can screw other fences to the plywood base for cutting angles.

1. Begin with a piece of ¾-in. plywood about the same size as the top of the tablesaw. The author prefers nine-ply Baltic birch plywood because it's so stable.

2. Make the runners from oak flooring, cut to be a little thinner than the depth of the miter-gauge slot. Mark #1 represents the full depth of the miter groove, while mark #2 is the thickness to cut the runner.

3. If you make the runners from a plainsawn board, their width ends up being quartersawn, which means the runners' dimension won't vary much with moisture changes. If the width of the runners did change, they would jam or become loose.

4. Epoxy will hold the runners to the plywood. Protect the top of the tablesaw with tape and place the runners in the slots. Shim the runners so their tops are flush with or slightly above the surface of the tablesaw.

5. Apply epoxy to the top of the runners and set the plywood on top. When the glue dries, add a couple of countersunk wood screws from below to ensure the runners stay put, and wax the runners so the sled slides easily.

6. A board attached to the front of the sled will act as a stiffener and a fence. Its alignment is critical. Crank the blade up through the plywood, stop it, and unplug the saw. Apply a coat of 5-minute epoxy to the bottom of the board, and align it with a framing square.

7. After the epoxy sets, reinforce the fence attachment with screws.

8. A stiffener added to the back of the sled keeps the plywood base rigid. Its alignment isn't critical, but don't make it too long or it can interfere with angle cuts on longer stock.

9. Because the fences are just screwed to the sleds, the crosscut jig can be used over and again to cut a variety of angles. What's important is aligning the fence at the proper angle to the blade.

10. Use a framing square to align a second miter fence.

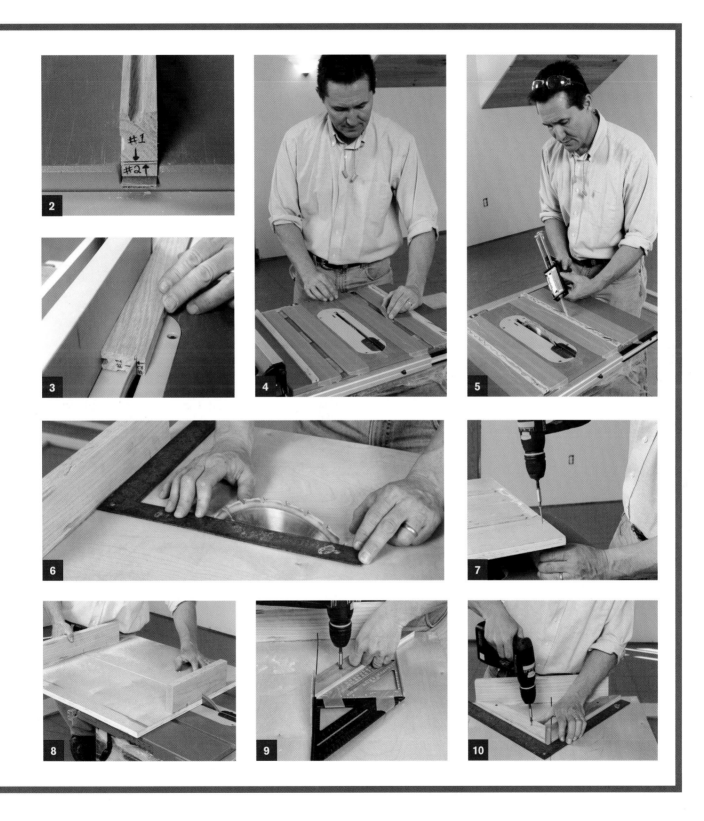

Micro-Adjusting a Fence

When cutting angles with a crosscut sled, the fence often needs the tiniest tweak to achieve perfect joints. I fine-tune by first setting the fence as close to the desired angle as possible and cutting test pieces. If adjustment is necessary, I'll shim behind the work piece at the front or back of the fence as appropriate, and cut more test pieces, adding shims until the cut is perfect. Dollar bills, which are about 0.004 in. thick, make great shims.

Once you've found the right number of dollar-bill shims, screw a block to the sled with the bills between the block and the fence. Unscrew that end of the fence, remove the bills, push the fence tight to the block, and screw the fence down using a new hole.

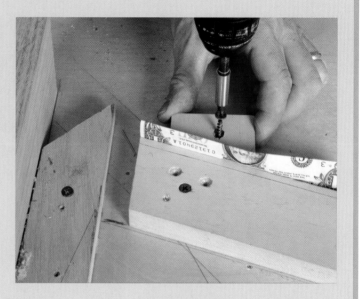

I usually rip the tongues from the flooring stock before cutting it into rhomboids, but not all installers do the same. Leaving the tongue on requires the extra step of grooving intersecting edges. In fact, you could groove all the edges and join the parquets with slip tongue. Because of the strength of urethane adhesives, I usually skip the slip tongue and just glue down the parquet.

The layout for the cubes is simple. I first mark my primary line and the perpendicular secondary line (as previously explained). I then dry-lay the parquet and mark my border working lines. The border working lines remind me where not to apply adhesive. The adhesive is stronger than the plywood so it is not easy to remove the cutoff pieces of parquet from the floor if you forget. On this floor, I had to shift the start of the parquets to the side because of the center medallion.

As with the herringbone floor, I mark the approximate border first. Unlike herringbone floors, I use just one layout line along the center of the longest axis of the floor. You could snap additional lines parallel to this first one, but I don't find it necessary. And forget perpendicular lines—they're meaningless in this case. I used no nails to install the cubes, just glue. At the border, the glue stops but the cubes run just past it to be trimmed later.

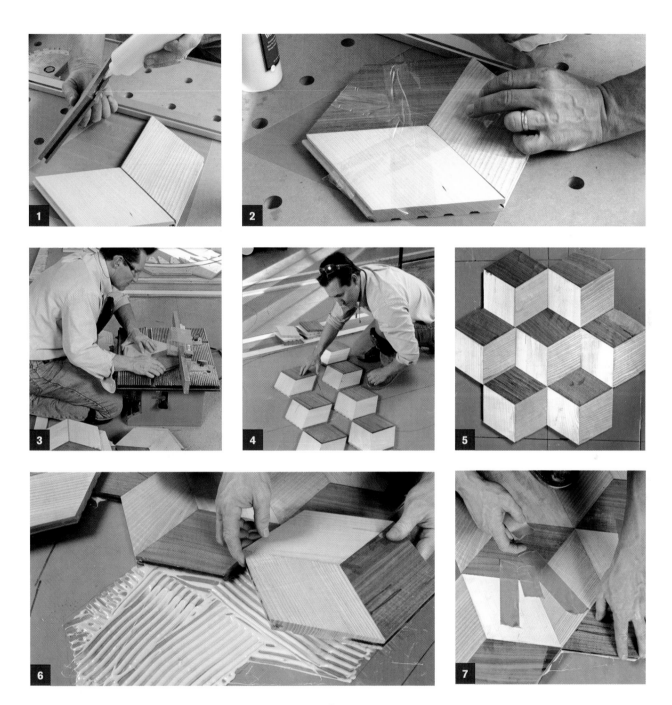

1. Glue holds the rhomboids together, but that's mainly to aid installation. Setting them in flooring adhesive creates the permanent bond.

2. Use tape to hold the sections together while the initial glue sets up.

3. If you do not cut the tongues off the original stock, you will have to groove one of the sides.

4. Dry-laying the rhomboid tiles helps you figure out the border working lines.

5. Layout of the rhomboids is relatively simple; they follow a centerline across the subfloor.

6, 7. The rhomboids are bedded in a moisture-cure urethane adhesive and held together with blue tape while the glue dries.

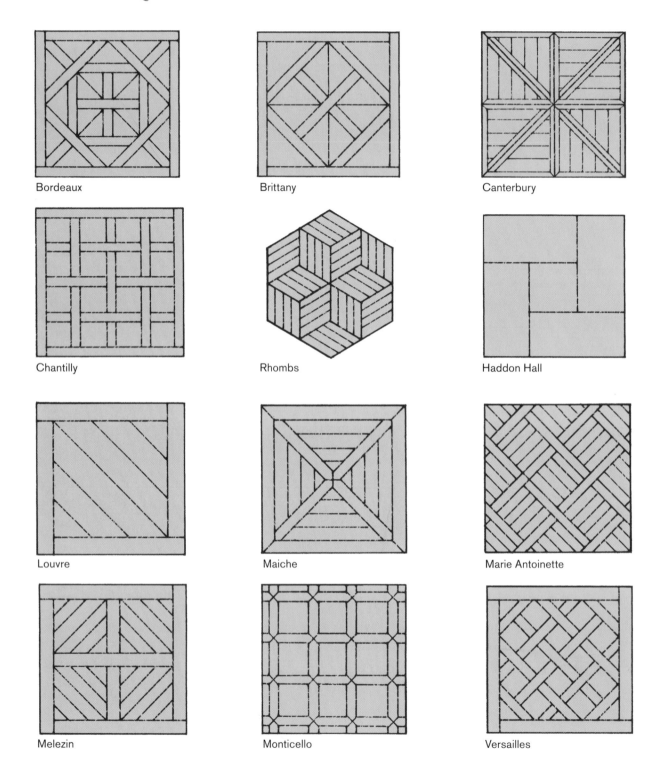

Bordeaux

Brittany

Canterbury

Chantilly

Rhombs

Haddon Hall

Louvre

Maiche

Marie Antoinette

Melezin

Monticello

Versailles

Square Parquet

Square parquet is available in many forms, including several patterns of glue-down tile, but with a little imagination you can make your own designs or copy historic floors (see the drawing on the facing page). Square parquets can fill a room wall to wall, or be surrounded by a decorative border. Additionally, inset individual tiles as accents in simpler floors.

MONTICELLO PARQUET

Thomas Jefferson at Monticello installed one of the earliest American parquets. These original parquets had a cherry center framed in beech; the parquets I used on the floor shown here have a center square of maple picture-framed with walnut. The Monticello pattern is relatively simple to make and to install.

The center of the parquet tiles shown in this section is a piece of 5-in. maple flooring. I don't like to go wider than that unless I'm using an engineered flooring product, because cross-grain expansion can open up the miters of the frame. I rip the tongue off all the material for the centers, both to get it out of the way and to ensure all the boards are of a uniform width. After ripping the center stock to width, I use a piece of it to set a stop block on the crosscut sled, and I cut all the centerpieces at one time.

Once I've cut the centerpieces, I use them to establish the length of the frame pieces. Using a crosscut sled with a 45° fence and a stop block, I cut all the picture frame pieces (see the photos on p. 150). To speed installation, I preassemble the parts into tiles using PVA carpenter's glue. I hold the pieces together with packing tape or blue masking tape until the PVA sets up. I don't really depend on this bond for anything more than keeping the parquet pieces together and making installation easier. What holds these, and most other parquets I make, together for the long term is the flooring adhesive that holds the parquet to the subfloor. Handle the freshly glued parquet with care and set it somewhere flat to dry. If you stack the tiles, put paper between them to prevent them from sticking together.

A picture frame of walnut surrounds a square maple block in this Monticello pattern parquet.

Engineered Parquet

To make parquet tiles from engineered wood, I re-saw ³⁄₄-in. flooring into ⁵⁄₁₆-in. veneers. I cut the veneers to the parquet pattern and then glue them to pieces of ¹⁄₂-in. Baltic birch plywood using Bostik's Best urethane adhesive. The combined thickness of these engineered parquets matches standard wood flooring, and I glue the assembled parquet pieces to the subfloor, again using Bostik's Best. PVA wood glues would also work to join the veneer to the Baltic birch plywood, but they would require a vacuum press bag to ensure the parquet pieces hold while the glue dries.

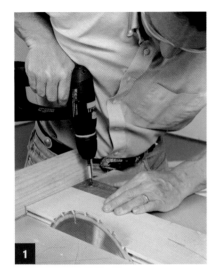

1. Picture frame parquet is simple to make. The key is to make the parts with a tablesaw sled set to make repetitive, accurate cuts. The center section needs to be a perfect square. Use the width of the stock to place a stop block at the correct width.

2. The crosscut sled precisely cuts the center pieces.

3. Set up with fences at 45° and a stop block, the sled miters the frame pieces perfectly. Fences on either side of the blade provide additional support for short pieces.

4. Use wood glue to join the parts. A great bond is not critical because the tiles will be fully adhered to the floor with moisture-cure adhesive. Scrap plywood screwed to a base forms a jig that aligns the pieces during assembly.

5. Use clear packing tape to hold the tile together until the glue sets up.

PAPERFACE PARQUET

Don't be confused by the terminology: "Paperface" parquet doesn't have a permanent paper face but just refers to the technique whereby more elaborate parquet tiles are held together before installation. By assembling small pieces of parquet in a form with no glue, they're easy to manipulate without getting glue everywhere. After arranging the pieces, you can roll on wallpaper paste and adhere a piece of heavy Kraft paper to the face of the tile. Now, you can pick it up, move it around, and

Making Paperface Parquet

Strips of $5/16$-in.-thick walnut are production-cut to make tiles. The walnut strips are glued to a paper backing, which is removed during installation.

An inexpensive combination square epoxied into a groove in the fence makes an adjustable stop for a variety of size pieces. With its blade secured, you can adjust the body of the square and then lock in place.

After cutting the initial miter, place the cut end of the veneer strip against the stop and trim it to length by sliding the sled back and forth over the tablesaw. Place different size pieces into separate boxes.

Once all the pieces are cut, roll starch-based wallpaper glue onto a square of Kraft paper placed inside a frame the size of the tiles.

Assemble the outer pieces first. Use no glue between them; the wallpaper paste on the paper temporarily holds the pieces together.

When assembly is complete, slide out the tile and set it aside to dry. The paper side is the top side of the parquet.

install it conveniently. You can make a surprisingly stunning floor from scraps of strip flooring. I have made a 500-sq.-ft. floor from scraps of $2\frac{1}{4}$-in. walnut flooring. To make better use of the material, I ripped the $3/4$-in. boards in half to make two $5/16$-in.-thick veneers. Then I sawed the veneer strips into two slats that were just less than 1 in. wide.

Because each parquet tile consisted of 48 separate pieces of wood (but only five different sizes), it is necessary to cut the slats with repeatable accuracy. As you can probably guess by now, I cut the pieces on a tablesaw using my crosscut sled. I also made up a prototype tile so I was

PAPERFACE PARQUET PATTERN

Composed of 48 separate pieces but only five different sizes, this parquet tile makes an elegant floor.

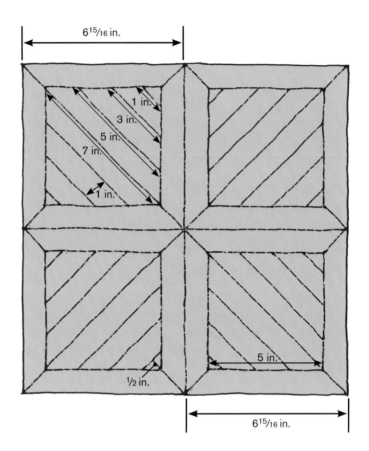

sure of the measurements. Because pieces of a variety of lengths were necessary, I made the sled with an adjustable stop. That's a fancy way to say that I epoxied the blade of an inexpensive combination square into a groove cut in the sled's fence. The body of the square can then move back and forth for different-length pieces.

It is easier to install fully assembled parquet squares rather than piecing the floor together on site. Using a jig the desired size of the tiles, I assembled them on a backing of 40-lb. Kraft paper using starch-based wallpaper glue. To assemble the parquet squares, it helps to make a frame to hold them. All it takes is three boards attached to your worktable. Leave one side open to slide out the finished parquet.

Paperface parquet installs like other square parquet, except that it's glued down with the paper side facing up, using a water-based PVA adhesive rather than a urethane. Immediately after setting the tile in the flooring adhesive, mist the paper face with water, to soften the wallpaper

It's hard to believe that this parquet floor was made using scraps from another job.

FAR LEFT After installing a few tiles, sparingly wet the paper with a sponge or spray bottle.

LEFT It takes about 20 seconds for the water to soften the glue enough to release the paper. Carefully remove the paper by pulling at a sharp angle and adjust any squares or individual pieces that are unaligned. Wipe off excess water on the parquet or the wood will distort.

paste, which allows the paper to be pulled away. The water in the PVA below the parquet balances the water on the face, helping to keep the parquet from warping. With the paper off, check the individual pieces for tightness and adjust as needed. Wipe off any excess water as you go. Any raised grain will be sanded smooth when the floor is finished (see chapter 10).

End-Grain Parquet

The patterns in most parquet flooring arise from some combination of mixing wood species and changing grain direction. Another approach is to glue down thin slices of wood so the end grain is the walking surface. When you look straight down at a tree stump, you are looking at the end grain. End-grain parquet floors are so durable they are used for industrial floors and even to pave streets. Similarly, butcher's cutting blocks take advantage of the slow-wearing nature of end-grain wood.

End-grain parquet can be any shape you want. The floor shown here is made from random sections of split logs. The layout for square or rectangular end-grain parquet isn't much different from any similar parquet, except that sometimes room is left between the parquets for grout.

If you install end-grain flooring in areas with great moisture swings, there are likely to be gaps. Because one edge will be quartersawn and

Although many end-grain floors are cut to exact tolerances like butcher blocks, these pieces were roughly cut from walnut log slices to fit together like fieldstone.

the other will be plainsawn, the sides of end-grain parquet change dimension at different rates. The plainsawn side will grow or shrink two times that of the quartersawn side. All you can do to minimize this is to acclimate the flooring before installation, use a vapor retarder below and a good finish above, and maintain the building's humidity within acceptable levels (see chapter 2).

I use urethane adhesive to install end-grain flooring. The rustic parquet shown here would have been too labor intensive to lay tightly, so I left grout space between the individual pieces. Because of how much end-grain flooring can move, I filled this space with a flexible grout made from ground cork and some water-based floor finish as a binder.

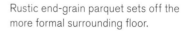

Rustic end-grain parquet sets off the more formal surrounding floor.

CHAPTER EIGHT

Inlays & Curves

UP TO NOW, WE'VE FOCUSED ON MASTERING THE straight and square realm of wood flooring, but when you introduce sinuous curves and inlays into your work, a whole new world of opportunity opens up. Whether the floor curves to meet a radiused wall or staircase, or curved forms are incorporated into the floor solely to please the eye, well-executed curves are truly the marks of a wood floor craftsman. These curves can be a part of the apron of a floor, or they can be decorative inlays in the field itself.

Not all inlays are curved, of course. They can be square medallions, stone tiles, or strips of metal such as brass, copper, or aluminum. What they all have in

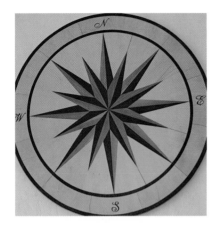

Stock inlays are available to purchase in a range of sizes, shapes, and wood species. Sources include: www.rare-earth-hardwoods.com, www.decorativeflooring.com, and www.globalinlays.com.

common is that the cutout in the flooring has to match the inlay precisely. The flooring is generally laid first and then carefully cut to fit using a router. This sounds tricky, but with the right jigs, it's actually quite approachable. Although you could fasten the inlay to the subfloor first, you'd then have to fit the rest of the flooring to it. This means that every piece of the intersecting flooring would have to be scribed to fit the inlay and then custom cut. Not only would this take much longer, it probably wouldn't come out as well.

Most inlays closely match the thickness of the flooring and are sanded flush before the floor is finished. Inlays can be purchased ready-made, but I prefer not to put my reputation on the line installing someone else's artistry. I like to make my own inlays from quartersawn $^5/_{16}$-in.-thick veneers glued to 11-ply $^1/_2$-in. Baltic birch plywood backer with a moisture-cure urethane adhesive. (You'll find more on these techniques in chapter 9.) The combination of quartersawn veneer and plywood backing makes for extremely stable inlays.

Begin with Layout

Laying out for inlays and curves follows most of the basic rules from earlier chapters—pay attention to focal points, bear in mind where furniture is likely to go, and so forth. Working off a centerline is usually the best approach, and, as always, you should be sure to verify the condition and moisture content of the subflooring.

Once you've decided on the main layout of the floor and marked it on the subfloor, mark the location of the inlays. This is important because the hole for the inlay will be routed into the flooring. Don't nail the flooring here, because hitting a nail will ruin a router bit. Secure the ends of the abutting flooring to the subfloor with a bead of construction adhesive, which sticks well to the trowel-applied urethane vapor retarder I prefer to use. (Note that the vapor retarder is missing in the top two photos on the facing page, taken at a class I was teaching.)

LEFT After deciding on the layout of the inlays, set them on the floor to verify how they look in place. When satisfied, mark their location.

ABOVE When laying the floor, extend the boards into the arc of the inlay, but don't nail at the cut line or beyond it. Use a bead of construction adhesive just outside the inlay to ensure the board ends stay put.

Circle Inlays

It may seem counterintuitive to anyone not accustomed to cutting curves, but a circle is probably the simplest shape to inlay. Cutting the hole with a router on a circle-cutting jig yields precise results. Commercial jigs are available, but I prefer to make my own, as explained in the sidebar on p. 160.

CUTTING OUT THE CIRCLE WITH A ROUTER AND JIG

To lay out the circle for the inlay, first set a pair of trammel points to the exact radius of the inlay (see the photos on p. 161). For the initial cut on the flooring, you want to set the circle-cutting jig a hair shorter than the radius. Be conservative here: You can always make the cutout larger, but you can't make it smaller. The pivot point of the jig attaches to a scrap block that screws to the subfloor. Once you make the first undersized cut-

Cutting circular inlays is simple.

Circle-Cutting Jig

Arc jigs for routers can be as simple as a router attached to a board that pivots on a nail driven through the far end, or somewhat more complicated like this adjustable circle-cutting jig. While there are commercial jigs available, I have not come across one that works better than this shop-made version. Pivoting on a wood screw, the length of this jig adjusts to cut an infinite range of arcs (photo at right).

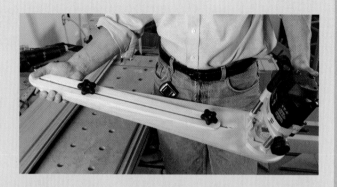

1. Start laying out the jig by tracing the base plate of the router onto $\frac{1}{2}$-in. Baltic birch plywood. Also, mark the location of the screw holes on the plywood at this time.

2. Lay out the two arms of the jig, which are a little more than 2 ft. long, so it can swing a maximum arc of about 4 ft. The piece that holds the router tapers to 2 in. at the end, which is the overall width of the piece with the pivot point.

3. Use a jigsaw to cut $\frac{5}{16}$-in. slots in each arm for the flat-head bolts, and then cut out each arm.

4. Drill and countersink holes in the plywood that align with those in the router base, and screw the two together.

5. Insert the $\frac{5}{16}$-in. flat-head bolts in the wider arm and secure the second arm with plastic knobs, which make tightening easy. Hardware is available from woodworking supply houses.

1. If you make your own medallions, using a circle-cutting jig and a router is the most accurate way to cut the floor. Set a pair of trammel points to the exact radius of the circle to be inlaid.

2, 3. Transfer the measurement to the circle jig, and make it slightly undersized for the initial cutout in the flooring.

4. After the initial cut, mark the exact circumference of the inlay on the flooring, and then set the circle jig and router to the radius. It may take several tries to get a perfect fit.

5. When making the final cut, start with the bit set to a depth of $1/16$ in. Such a shallow initial pass minimizes tearout.

6. Affix the inlay permanently with moisture-cure adhesive.

out, mark the exact circumference of the inlay on the flooring, and then set the circle jig and router to the radius (half the diameter). When making the final cut, start with the bit set to a shallow depth of about $\frac{1}{16}$ in. to minimize tearout. Check the fit of the inlay in the cutout, and then apply moisture-cure adhesive to the subfloor to affix the inlay permanently. If you have access from below, you can run short screws through the subfloor into the inlay to reinforce the joint. Be sure to check the screw length first.

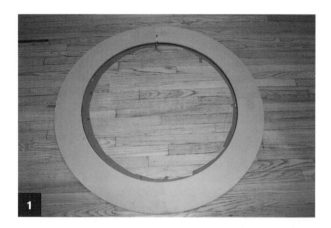

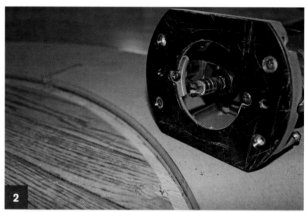

1. Most purchased medallions come with a template for guiding a pattern-routing bit. Use double-sided tape to hold the template to a finished floor.

2, 3. Drill a hole in the floor to provide a safe place to start the router bit, and make the first pass shallow to minimize splintering.

4. It's sometimes impossible to avoid all nails. If you encounter one (the author ran into seven on this floor), dig down to it with a chisel and drive it down with a nail set and hammer or pull out with a pair of locking pliers.

USING A TEMPLATE

If you're inlaying a purchased medallion rather than one you've made yourself, it will likely come with a template to guide the router to make the cutout in the flooring (see the top left photo on the facing page). Be sure to first check the area of the cutout for existing nails (which would damage the router bit) by using a magnet. Usually, you can avoid nails by shifting the position of the medallion slightly—no one will notice if it is ½ in. off center. To secure the template to the floor, use either double-faced

5. Once the cutout is routed to the full depth, remove the boards with a pry bar and hammer.

6. Uneven floors require flattening. An edger makes quick work of high spots.

7. Set in a bed of moisture-cure adhesive, the medallion is held down while the glue cures with either screws from below or weight from above.

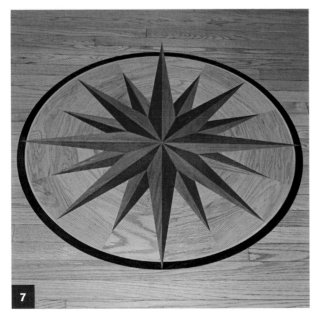

Oval Medallions

Unlike round medallions, oval ones are difficult to inlay using a router jig. Oval-cutting jigs are complicated affairs with guides that slide in grooves oriented as *x* and *y* axes. Such a jig fitted with a router can create the oval-shaped medallions, but the jig is not accurate enough to create a duplicate hole in the floor. Normally, flooring installers will make the oval medallion and trace its shape onto a piece of plywood, which is cut with a jigsaw to make an installation template. It may require hand sanding of high spots, or a layer of tape to fill in low spots to make it just right. Happily, you can purchase oval medallions that come with templates for use with a pattern-cutting bit.

tape or finish nails driven into the flooring where you will install the medallion. If you're using finish nails, aim for the joints between the boards so that you don't damage the finished floor. The template guides a ¾-in. pattern routing bit in cutting the flooring, and a bearing placed just above the cutter of the bit matches the cutter's diameter. The bearing rides along the surface of the template, and the cutter exactly reproduces its shape in the surface below.

Purchased medallions are available in thicknesses from thin-engineered flooring up to standard ¾-in. flooring thickness. It's crucial to specify the exact thickness if the inlay will be installed flush in an existing floor that won't be sanded and finished. To determine the required thickness, I drill a hole in the area where the medallion will go using a ½-in. or so spade bit. Use the kind with spurs and drill slowly. As soon as the spurs have cut completely through, the remaining flooring will start to spin. Stop drilling and remove the plug. Now you can measure from the top of the flooring to the top of the subflooring with a machinist's depth gauge (a thin, flexible rule with a small sliding T). Subtract from this the thickness of a piece of #15 building paper, about 0.015 in., to allow for the glue. Determining the exact thickness isn't as important if the entire floor is to be sanded and finished, though it goes without saying that the medallion should not be thinner than the floor in which it's to be installed.

STAR-SHAPED MEDALLIONS

Star-shaped medallions are a little more difficult to install than circles and squares, but easier than oval shapes. There is no simple way to cut all the points of the star into the floor. You can make a template of the star from plywood, but the easiest way to obtain an accurate cut is to set the star in place and surround it with small pieces of wood double face taped to the floor. Using a pattern bit in a router will cut everything except the very tips of the star. These can be finished with a sharp chisel, corner chisel, or a Fein Multimaster® (www.fein.de/com).

To inlay stars or other complicated shapes, make a router template by holding strips of wood tight to the medallion and fastening them to the floor.

Laminated Curves

My favorite curves are ones in which the wood grain follows the bend of a wall or other architectural element. They look as though cut from a forest of bowed trees. You can achieve this look by cutting a board into thin strips and gluing them back together against a curved form. Once the glue cures, the strips will hold the curved shape, though with some "springback." Springback occurs when the lamination releases from the form and the tension in the bent wood pulls it slightly back toward being straight. Springback is usually minor enough that the inlay can still be installed. The tighter the radius and the fewer and thicker the strips, the more springback you can expect. Of course, using thinner strips means that you need to use more strips, which complicates matters at glue-up. To compensate for springback, you can form the lamination at a slightly tighter radius, but I rarely find this necessary.

Hardwoods generally bend better than softwoods (see the sidebar on p. 166), and strips ripped from quartersawn stock bend better than those ripped from plainsawn boards (see chapter 1). Bending stock should be free of weak areas like knots, distorted wood grain, shake, pith, and surface checks. Boards generally cannot be bent with a radius less than 20 to 30 times the wood's thickness. So, for example, the most flexible

The flowing grain of bent-wood laminations follows and accentuates architectural elements such as a bullnose step.

The Best Woods for Bending

Some woods bend better than others. The USDA's Forest Products Laboratory in Wisconsin researches and publishes a wealth of information on U.S.-grown wood. The list at right shows common American wood species in descending order of suitability for bending—in other words, white oak is the most suitable American wood for bending and mahogany the least suitable. Unfortunately, I know of no similar list for imported woods, though I can say from experience that some exotic hardwoods such as Brazilian cherry can be brittle and have more of a tendency to break during bending. That said, I've bent Brazilian cherry many times. One of the tricks is to cut the laminations from quartersawn boards.

White oak

Red oak

Pecan

Black walnut

Hickory

Beech

American elm

Birch

Ash

Soft maple

Yellow poplar

Hard maple

Chestnut

Mahogany

Radial face

Tangential face

Most wood species bend better when using quartersawn stock, with the radial face up. Additionally, use the straightest grain boards you can find. Using such stock reduces the chances of cracking and can make it possible to bend even relatively brittle species.

¼-in.-thick laminations would bend at most to a radius of 5 in. More brittle woods of the same thickness might bend only to 7½ in.

In the project shown on pp. 169–171, I ripped the strips from maple flooring. If you work with strips of flooring, you have to be careful during glue-up. Because flooring is grooved on the bottom, some of the strips you rip will likely be a little narrower than others. You have to pay attention that they don't slip below the tops of the others. Another approach is to rip the strips from 5/4 lumber, laminate the curve, and, once the glue is cured, run the lamination through a thickness planer. I didn't use 5/4 for this lamination because I was using quartersawn maple flooring, and I couldn't match that in 5/4. So I used ¾-in. flooring, and was very careful to keep the tops aligned.

Making the Bending Form for a Laminated Curve

A laminated curve is typically just one element in a larger ornate floor. In the floor shown here, which you have seen in a number of chapters, the laminated curves feature at either end of the central panel, where they contain the three-dimensional parquet floor that we highlighted in chapter 7. The radius of the curving border was determined after installing the field parquet. From that point, creating a laminated curve is a matter of making the form, ripping and gluing strips, and cutting out the flooring.

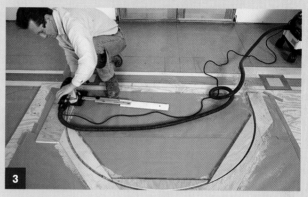

1. Make the bending form on a sheet of plywood covered with Kraft paper. First, mark the curve's radius on the paper to provide a guide for the form, then cut and arrange strips of ¾-in. plywood to cover the curve.

2. Screw the ¾-in. form boards to the plywood. Use trammel points to transfer the arc to the tops of the form boards.

3. Set a router and a circle-cutting jig to cut the outer radius on the bending form.

LAYING OUT THE CURVE

As with any ornate flooring, the first step is layout. Decide where you want the curve, and what its radius will be, then draw it out on the floor before installing the field flooring. This way, you can be sure to avoid putting glue or fasteners in areas that will be cut out for the inlay. Once you've laid out the floor, you can proceed to laminating.

I like to use common PVA carpenter's glue such as Titebond® II for the laminations. Although it has an open time of only about 5 minutes

PVA vs. Polyurethane Glue

Some installers prefer to use poly-urethane glue for the laminations, but I don't like the way it foams up, and there can be no gaps between the laminations. PVA glue is more forgiving in that it will bridge small gaps. One advantage polyurethane glue does offer is a longer open time than PVA, which can be help-ful. This is balanced somewhat by the fact that polyurethane must be cleaned up with mineral spirits, and then only before it dries. If it gets on your fingers, the stain from it stays there until the skin wears away. PVA glue cleans up with water even after it dries (as long as you don't use the water-proof varieties).

You can tint PVA glue with aniline dyes to match the wood. A glue roller aids in quickly applying an even coating to the laminations.

and an assembly time of 10 to 15 minutes, this is generally long enough. (For speed, use a roller or flat trowel to spread the glue.) Titebond II dries to a translucent yellow color, but you can mix aniline dyes such as TransTint® (Homestead Finishing, www.homesteadfinishingproducts.com) with the glue to help match the color of the wood laminations. That said, nearly any wood glue will work for laminations, as they will be fully glued to the subfloor anyway.

No matter what glue you choose, use enough clamps to bring the laminations tightly together; and it's not a bad idea to do a dry run first. Because of the tension inherent in laminated curves, it's important to keep them clamped for the maximum time suggested by the glue manufacturer. With PVA glues, that's usually 24 hours.

The strips for the laminations should be long enough to make the bend in one piece, and then some. You can determine the length by measuring along the bending form with a tape measure, or mathematically. The circumference of a circle is pi times the diameter. If you're doing a half circle, as shown here, divide the circumference by two. Also, keep in mind that if the laminations measure, say, $\frac{1}{8}$ in., the $\frac{1}{8}$-in. kerf of a tablesaw blade means you'll have a 50% waste factor. Thinner strips waste even more, thicker strips less.

Deciding on the thickness and number of strips to use in a laminated curve is always a judgment call based on the kind of wood, the radius, acceptable waste, and time. Not all species of wood bend as well as others (see the sidebar on p. 166). When using flexible species, you can use fewer and thicker laminations, that is, closer to $\frac{1}{20}$ of the radius. With less flexible species, laminations that are around $\frac{1}{30}$ of the radius are in order. It's always easier to get thinner laminations to bend, so why not just use more and thinner strips as a rule? There's additional waste when ripping thinner strips, of course, but time becomes the bigger factor. Every glue has a limited open time, the period you have after spreading the glue to clamp the assembly. The more strips in a lamination, the more of that open time you use spreading the glue.

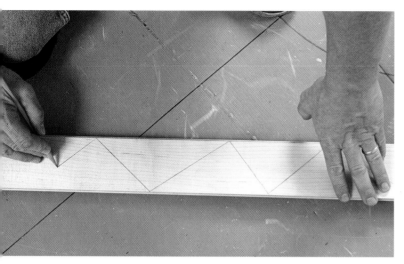

1. Mark the maple board before ripping it into strips. The marks help align the laminations so the grain of the curve will look more natural if assembled in order.

2. A tablesaw cuts the strips precisely, but bear in mind that the kerf can waste half of your stock. A band saw's thinner blade wastes less, and preserves the grain pattern better.

3. Verify that the strips are thin enough to bend without breaking. The radius here is about 2½ ft., and to make the required 2⅝ -in. wide laminate, 14 maple strips ³⁄₁₆ in. thick were used. Make the strips longer than the curve to allow for trimming.

1. Because of the glue's short (10-minute) open time, speed is essential. Use a roller or flat trowel to apply the glue quickly.

2. Pile the glued-up strips, match up the alignment marks, and place the bundle against the form. Use strips of Masonite® between the clamps and the lamination to avoid denting the lamination and to spread the clamp pressure. Be careful not to glue these strips in place.

3, 4. Shop-made plywood cams screwed to the working surface make for fast clamping. Start in the center and work toward the ends. Blue tape applied to the form prevents the lamination from attaching permanently to it.

5. Wedges help tighten stubborn areas.

INSTALLING THE INLAY

The key to installing any curved inlay is to cut away the flooring accurately. I've said it before, but it bears repeating that you need to lay out the curve on the subfloor before installing the flooring. This way, you can avoid nailing in areas that you'll cut out. Also, the remaining ends of the boards to be cut should be glued fast to the subfloor, and the layout lines show you where not to put the glue.

Install curves with a simple arc using a router arc jig. For complex curves, trace the completed arc to a piece of plywood first to make an installation jig. Cut out the plywood with a jigsaw; it may require a little hand sanding or a layer of tape for a shim to make it just right.

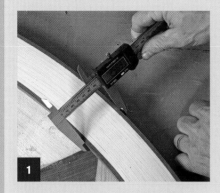

Installing a Laminated Curve

Once you've made the laminated curve, you're ready to trim it to length and install it into the floor. To do this, you'll need to rout out a curve to accommodate the laminate. It can take a little courage to rout out a curve in a floor you've labored to install, but using a sturdy router jig and cutting conservatively removes most risk. By cutting conservatively, I mean you should make the first cuts just inside the layout lines, and extend them incrementally so you sneak up on a perfect fit.

1. Measure the width of the completed lamination. This tells you (a) the width to rip any adjoining straight stock and (b) how far to move the router in or out from the first cut.

2. Kraft paper and a wooden block glued to the wood floor provide a removable pivot point for the router jig without making a hole in the floor. Carefully center a pivot hole in the block by marking the center of the field, and then measuring the radius from where you decided the tip of the curve should go.

3. The radius of the curve was marked on the floor with trammel points to provide a reference for the router jig, and to check that the laminate will fit before making the cut. To play it safe, cut the opening a little undersize and then check the fit of the curve. It's better to have to widen the area with a skim cut than to cut it too big.

4. Test-fit the curve. Slight bumps in the lamination may require fairing with a sanding block. Use a long straightedge square to the border to mark the end cuts on the lamination.

5. Apply moisture-cure urethane adhesive to secure the curved inlay to the subfloor.

Stone inlays add elegance and can help link wood floors to their stone counterparts elsewhere in a house.

Stone and Metal Inlays

All sorts of materials can be inlaid in a wood floor, including stone and tile, various metals, glass, and leather. Such decorative elements can serve to jazz up a plain floor, highlight a focal point, or, by repeating the use of a material, tie together the design of a whole house. Most of the materials people inlay are at least as hard as the wood flooring, so you want to avoid sanding them. Metals, for example, are usually installed just below the surface of the raw flooring, which is then sanded flush. (With soft metals such as brass, copper, or aluminum, brads are sometimes used to secure the inlay until the glue cures. The heads of these are small enough and soft enough that sanding them flush is not a problem.) Stone and tile are not sanded at all, but rather are set after the floor is finished.

INLAYING STONE

Stone inlays are generally made from commercially available tile, which runs around ⅜ in. thick. To bring the tile flush with the flooring, mount it to plywood backing with urethane adhesive. You can install thicker stone, but it should not be thicker than the flooring. Because stone and tile don't expand and contract with changes in humidity, these inlays require clearance to the wood flooring, which usually ranges from ⅛ in. to ¼ in. Fill the gap with caulk the color of the stone after gluing the plywood/stone lamination to the subfloor.

INLAYING METAL

Soft metals such as brass, aluminum, and copper can be inlayed in wood floors (but stay away from ferrous metals that can rust). There's no reason for the metal to go all the way to the subfloor, so metal inlay material is generally ¼ in. thick. Best practice is to install the metal with adhesive and nails made of the same metal. Two-part epoxy, moisture-cure urethane adhesive, and polyurethane all work. I set metal inlays $\frac{1}{32}$ in. below the sanded wood floor surface, as it's easier to sand the wood flooring down to the metal inlay than vice versa. Designs should take

1. Attach the stone veneer to Baltic birch plywood backing with urethane adhesive.

2. To avoid scratching the stone face, place the inlay upside down while sanding the floor. Adding two pieces of #15 building paper (or a credit card) below the stone emulates the thickness of the adhesive, so when installed the stone will be flush with the wood floor.

3. Remove the paper shim and secure the stone inlay with urethane adhesive.

Inlaying Metal Bar Stock

Readily available brass bar stock, usually ¼ in. sq., is a simple way to add a unique look to a floor. Most carpenters already own the tools, and any additional tools aren't expensive to purchase. Epoxy and polyurethane adhesives work well for installing metal inlays. Nails of matching material are generally installed to back up the adhesive.

1. Cut the metal stock to size and drill nail holes every 9 to 10 in.

2. Clean the metal with a solvent and an abrasive to ensure a good adhesive bond.

3. Use polyurethane glue to adhere the brass stock to the flooring.

4. Nails every 9 in. hold the brass in place. Set it slightly below the surface so you're sanding the wood to meet the metal.

into consideration that metal inlays will not change dimensionally with wood during swings in humidity.

A particularly elegant touch is to inlay curved metal strips. To many carpenters and woodworkers, the idea of curving metal is even more intimidating than curving wood, but with the right tools, it's actually quite simple. With a circle-cutting jig for a router to cut the floor, and a planetary ring roller for the metal bar stock (about $80 as of this writing), you can trick out a wood floor in a way that will really wow the audience.

A somewhat whimsical approach is to rout a series of grooves that are inlaid with brass or contrasting wood. Metal inlay can even ease

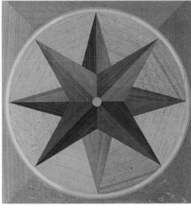

A brass inlay encircling a medallion sets it off with style.

1. Using a ¼-in. straight bit and a circle-cutting jig, cut a groove slightly more than 1/32 in. deeper than the thickness of the brass to compensate for the adhesive and to set the brass slightly below the surface of the flooring.

2. A planetary ring roller bends metal stock to a uniform curve. It adjusts for various radii.

3. Depending on the length of the metal, make the ring in segments. Several passes are necessary for tighter radii. Check the diameter of the ring segment as you go.

4. Use a rotary tool with a cutoff wheel or a hacksaw to trim the brass segment to size.

5. Brass nails and 5-minute epoxy hold the ring in place. When sanded, the nails will blend into the brass ring.

your task when creating round medallions. It's more difficult to make a medallion where all the wedge-shaped segments come together perfectly in the center. But if you inlay a brass button at the center, it not only adds a point of emphasis, it also makes the medallion look as though it was harder to make when, in fact, it became easier.

CHAPTER NINE

Making Medallions

AS DISCUSSED IN CHAPTER 8, MEDALLIONS and other inlays are decorative elements normally incorporated as the centerpiece and focal point of the floor (though they can also be used in borders). In the previous chapter, I explained how to install manufactured medallions and other inlays, but here I'm going to look at the various options for making your own medallions.

There are many reasons why I prefer to make my own medallions. Manufactured medallions are generally cut by a computer-controlled laser or router, and some manufacturers use lower-powered lasers that can leave black burn marks on the edges of each

TOP Inlays can comprise a variety of elements, including the medallions and scrollwork shown here.

ABOVE This manufactured medallion developed large gaps after installation because it was improperly acclimated prior to assembly. The pie-shaped piece of maple in the center is fractured and should have been culled. You should always carefully inspect inlays on delivery.

piece. The wood used in some medallions may contain wood colors and marks that you might not select if making the medallion yourself. And, finally, the medallion may have been made in an environment that differs from your home. For example, you may be installing a floor in Tucson with its desert humidity. A medallion made in the humid south-east could shrink and crack dramatically.

I get more satisfaction from installing medallions I've made than from the ones I buy, and if a customer has a specific design in mind, it's likely I can produce it. Unlike the manufactured medallions I've seen, I create mine only from stable rift-sawn and quartersawn material and acclimate the wood prior to assembly. In this way, the quality of the job I do doesn't depend on anyone else making a high-quality medallion.

Inlays take many forms beyond medallions. Simple inlays—strips or pieces of stone, for example—don't need to be made, and they were covered in the previous chapter. Another kind of inlay, scrollwork, where intricate shapes mate together, is covered at the end of this chapter. Technically, though I refer to all decorative elements as inlays, true inlays are fitted just into the surface of the wood flooring, and not down to the subfloor.

Medallion Material

Medallions can be made from full-thickness flooring (like the St. Mark's medallion described on pp. 185–189) or from veneers affixed to plywood (as most of the other medallions in this chapter are done). If you use full-thickness stock, make sure it's quartersawn for stability and permanently glue the medallion to the floor as soon as you can to minimize the chances of it warping and splitting. If the inlay is round, use only more stable woods. Hickory and beech move more than oak, which moves more than cherry and alder. Quartersawn stock of any species moves about half as much, and often less, than plainsawn stock (see chapter 1).

Round medallions are typically made by assembling a bunch of wedges into a circle, which means the entire circumference of the circle will expand and contract like one piece of cross-grain wood. Since the tops of the wedges are wider, they change dimension more with humidity swings than the tips. Consequently, they have a tendency to break apart with slight changes in moisture content.

Because of the difficulty in keeping solid medallions from cracking, I make most of mine from 5/16-in.-thick quartersawn veneers glued to 1/2-in., 11-ply Baltic birch plywood (see the top right photo on p. 180). Since it doesn't change dimension much with variations in humidity, the plywood makes for more stable medallions. I slice the veneer from 3/4-in. flooring, getting two pieces out of each board. The veneer is thin enough not to exert much stress on the glue joint, so the plywood is able to keep the veneer from moving significantly. I glue the veneer to the plywood with moisture-cure urethane adhesive. PVA glue would also work, but

This starburst medallion takes advantage of and emphasizes the unique shape of the room.

it would require something like a vacuum press bag to ensure the veneer pieces are fully held down while the adhesive dries. Additionally, the bond strength of moisture-cure urethane adhesives does not degrade over moisture cycles as other adhesives can.

Medallion Shapes

The most common shapes for medallions are squares, circles, and stars. Square and round medallions are the easiest to make and install. Because of the intricate cutout in the floor they require, star-shaped medallions are harder to install. It's much easier to create and install a round medallion with a star-shaped center.

No matter what shape medallion I'm making, it's a good idea to draw it out full scale first to provide a template. Because of the number of angles and individual parts in most medallions, accuracy is crucial. The smallest error repeated adds up to unacceptable gaps, and a full-size drawing provides a ready visual reference. Trammel points—essentially, clamps that hold a pivot point and a pencil to a length of wood—serve as a large compass. I use them not only to draw circles but also to construct angles.

Medallion design is frequently geometric. Starbursts and compass roses are a common design. Not only do they look good, but also their radial layout lends itself to being fit in a round medallion. You shouldn't feel limited to a strict geometric pattern, however. Marquetry woven into the centers of medallions can yield stunning results.

Installing a star-shaped medallion requires painstaking work to cut out the flooring. The cutout for a round medallion is far easier to do, using a router and a circle-cutting jig.

FAR LEFT This 12-sided medallion isn't the easiest to install, but its dynamic geometric design certainly catches the eye.

LEFT The compass rose is one of the most traditional designs.

Dividing a Circle

Often, medallions are composed of a series of wedges that combine to fill a circle. Although you can determine the angles of the wedges mathematically (360°/number of segments), it's helpful to draw and accurately divide the circle to visualize and determine the wedge sizes.

In addition to 90° angles (as shown in chapter 7), trammels can lay out end points for lines that accurately bisect them into smaller angles. To do this, lay out intersecting arcs centered on where square lines intersect the circle (photo below left).

A line connecting two pairs of intersecting arcs divides the square into two 45° wedges (photo below right). These can be further broken down into 22.5°, 11.25°, and so on.

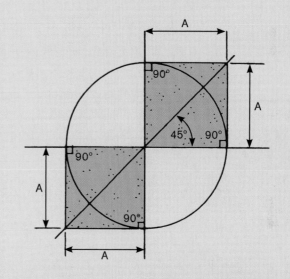

Although the 32-piece medallion shown here looks complicated, it uses only four different-size pieces of wood. Once you've built a prototype, this makes producing duplicates a matter of setting up stops on a tablesaw sled and production cutting parts. To bring the medallion to life, the pieces are from several species of wood—red oak, cherry, and Brazilian cherry.

Square Medallions

Square medallions are the easiest to make and install. You don't need special jigs to cut out the flooring, though they can make it easier. The angles of the medallion pieces tend to be ones people are comfortable with—90°, 45°, 22½°. This helps because it's easier for most people to envision, say, a 45° angle than a 15° angle. To make them even easier to install, keep the overall size of a square medallion a multiple of the flooring board's widths so it fits without having to notch the flooring. Including an outside feature strip of a contrasting wood both helps to set off the medallion visually and provides an area that can be adjusted in width so it fits within the floor's layout.

Square medallions are relatively stable and can be made from ¾-in. stock with little risk of problems due to seasonal expansion and contraction. Draw the medallion out full scale on plywood so there's no guessing about the sizes of any pieces.

Medallions usually consist of a number of small pieces, and cutting them accurately and safely requires care. I use a crosscut sled on the tablesaw for almost all the cuts, which allows me to make repeated cuts with precision. The sled also aids in holding the small pieces so I keep my fingers out of the blade. For any cuts that feel risky, I clamp the workpiece down.

Begin ripping the stock to width. Cutting the center wedges is the trickiest part of the medallion as the pieces are relatively short. Cut these in a two-step process (see the photos on p. 184). To be safe, make the first angled cut from longer stock. The second angled cut detaches the short piece from the longer one. The cuts for the arms of the star are more straightforward. Check the pieces for fit, and then glue them together before installation.

(continued on p. 185)

Change the Wood, Change the Look

Endless variations of square medallions are possible. Additional feature strips can be added, different woods used, or the inner design can be changed, as here where the arms of the star were cut from Brazilian cherry and white oak to alter the center. Boards of each species were glued together before being cut to size and shape.

When using multiple medallions in a floor, or throughout the house, one approach is to use variations of the same design. This provides visual interest, while maintaining a rhythm to the design.

PATTERN FOR SQUARE MEDALLION

Dividing this square medallion into eight segments limits all the cuts to 45°. Alternatively, the center could be composed of an intricate star.

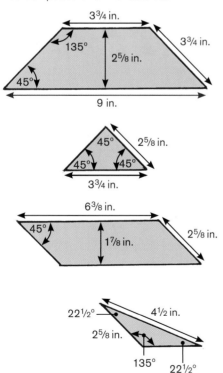

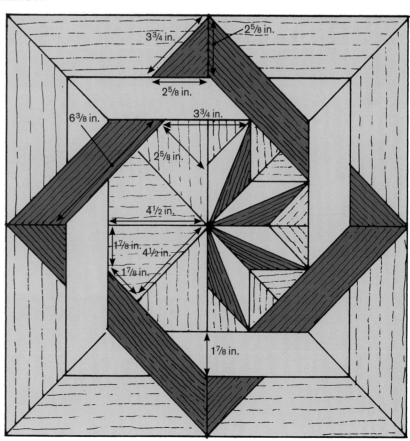

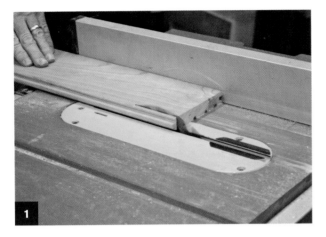

1. Uniform stock is a key component. Rip as much as you'll need of each width to avoid having to try to reset the fence to cut another piece of a particular width stock.

2. To cut the center wedges, align a square-cut board so that a 45° cut yields a side $1\frac{7}{8}$ in. long.

3. Flip the stock and cut at 90° to the first cut, preserving its $1\frac{7}{8}$ in. length.

4. Cutting the longer arms is more straightforward than cutting the center wedges.

5. Check the fit of medallion pieces prior to assembly. Once you're sure of the fit, join the pieces with carpenter's glue and hold them together with blue tape until the glue cures.

(continued from p. 182)

If desired, you can hold the medallion together temporarily with tape instead of glue, though this makes installation more difficult.

Round Medallions

Curves are pleasing to the eye and convey a sense of craftsmanship. Round medallions aren't hard to make and install if you have a router and basic arc jig (see chapter 8). As with square medallions, I start out with a full-size drawing to aid in sizing the parts.

The construction of all the medallions shown in this section is similar in that their centers are some form of wedges that join together to form a full circle. The medallions are put together so that the outside of the wedges run long and then the entire assembly is trimmed to a circle using a router and a circle-cutting jig. Once you master the techniques for parts, you can build nearly any medallion imaginable. Let's begin with a simple design, what I call the St. Mark's medallion because it reminds me of the marble floor medallions in St. Mark's cathedral in Venice.

ST. MARK'S MEDALLION

I made the St. Mark's medallion shown on p. 179 from solid ¾-in. flooring, but it could also be made from ⁵⁄₁₆-in.-thick veneer glued to ½-in. plywood backing. Finding the angle of the apex is simple enough: 360° divided by the number of wedges. The St. Mark's medallion uses 12 wedges each made from two triangles of walnut and cherry for 24 pieces (see the drawing on p. 186). Capping the assembled wedge is a maple triangle.

CUTTING THE WEDGES To find the angle of the wedges, divide 360° by 24. The result is 15°, and that's the setting for the fence on the sled. To set up a template at the correct angle, I use a simple trig function, as explained in the sidebar on p. 190. Next, cut a full complement of wedges from scrap, and see how they go together. It's likely the fence will need a slight adjustment (see the top right photo on the facing page). You can measure any gap left after assembling the wedges and divide it by the number of triangles to get an idea of how much you need to adjust the fence. Use mechanic's feeler gauges or a dollar bill to measure the adjustment (see top right photo on p. 187). You may need to cut several trial sets before

Round medallions are eye-catching, but they aren't as difficult to make as they appear. The St. Mark's medallion shown here has 36 triangular-shaped pieces, plus a round border.

WEDGES FOR THE ST. MARK'S MEDALLION

Two wedges of contrasting woods make up each segment.

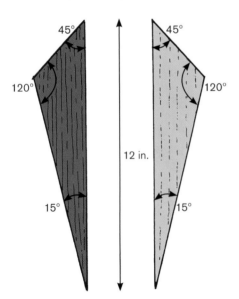

you get the setup perfect. Be sure to set this primary fence far enough from the end of the sled that the triangles you cut will be 1 in. or so longer than the radius of the circle.

For the wedges used in the medallion to seat against the end of the sled, first cut the end of the workpiece to 15° so it seats against the end of the sled. This requires a secondary fence to make the cut. Once the ends are properly angled, cut the triangles. The wedges of one species are cut good-side up, and the wedges of the other are cut upside down. This wedge-cutting jig is used for several of the round medallions in this chapter. In fact, because of the number of medallions I build that use this initial setup, I leave this fence in place on a dedicated crosscut sled.

After the initial setup of the cutting jig, cut a full circle's complement of wedges and test the fit. They should make a perfect circle. The next step is to set up a crosscut sled to cut the angle on the top of the triangles. In this case, I wanted the tops of the triangles to angle back 45°. This angle is not critical for the St. Mark's medallion, but it is for the starburst medallion shown later in this chapter (see pp. 191, 196–198). In the interest of efficiency, I also set up the St. Mark's jig to cut the starburst pieces. You could use a different angle if you aren't making the starburst medallion.

A secondary fence on the jig for cutting the tops of the triangles controls their length. To set this up, first mark the desired length on the triangle, and then hold it in place for a cut, and screw down the secondary fence. Cut the walnut and cherry triangles with one species right-side up and the other upside down so they mirror each other when assembled.

CUTTING THE CAPS Once all the cherry and walnut triangles are cut to size, set up another fence on the sled to cut the maple caps. This fence should be at 60° to the blade and is set by using a cutoff from the triangles. The maple needs to be wide enough so that, when assembled, the center of the cap is at least the radius of the medallion from the points of

(continued on p. 189)

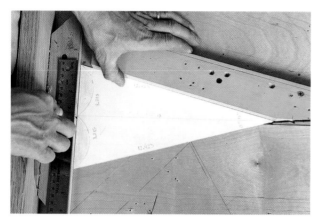

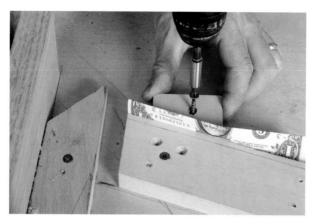

1. Use a cardboard template to set the fence on the sled initially. Use this setup to cut 24 trial wedges. With the wedges assembled into a circle, any gaps indicate an unaligned fence.

2. A dollar bill is 0.004 in. thick and makes a handy shim for adjusting the fence (if necessary, fold to increase thickness). Place the shim against the fence, press a temporary block against it, and screw the block down. Loosen the fence, remove the shim, press the fence against the block, and refasten it.

3. Once the primary fence is perfectly aligned, align a secondary fence (with the clamp on it) square to the primary fence. Use the secondary fence to trim the wedge ends to 15° so they seat against the end of the sled.

4. Clamp the workpiece so it's tight to the fence and the end of the sled and cut it to 15°. Be sure to clear away any sawdust that would prevent full contact between the work and the fences.

5. Cut the wedges long, test fit them in a circle, and then mark them to length. In this case, that's 12 in.

FAR LEFT Set the fence closest to the operator at 60° to the sawblade (to yield the required 120° cut). Set the fence away from the operator so that when the workpiece seats fully, it's set for cutting at the proper length. (Here, the cherry wedge is being cut.)

LEFT Flip the walnut wedge upside down and cut it as a mirror image of the cherry wedge.

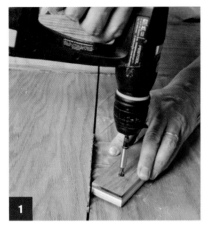

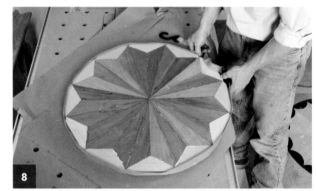

1. Use an offcut from the wedges to set the fence angle for cutting the maple caps.

2. Set up a secondary fence with a clamp to hold the maple pieces.

3. Flip the maple piece over to duplicate the angle.

4. Verify the fit of the cap with the cherry and walnut triangles.

5. Mark the tips of the maple pieces to be trimmed.

6. Add a stop to the jig to trim the caps to a uniform length.

7. Apply wood glue with a roller. Work fast, as the glue's open time is only about 10 minutes (less in hot, dry conditions).

8. Use a ratcheting strap to clamp the medallion while the glue cures.

(continued from p. 186)

the walnut and cherry triangles. When trimming the assembled medallion to a circle with a router, the circumference will arc through the maple caps between the tips of the wedges. This circumferential cut will trim the ends of the maple caps flush with the tips of the wedges. For this to work, the center of the triangular maple caps must extend past the ends of the walnut and cherry triangles.

Set up a secondary fence on the sled so the maple stock fits tightly between it and the primary fence. The maple pieces will be too small to hold safely by hand, so you'll need to set up some sort of clamp with that fence. Cut the caps at reflexive angles by flipping the stock over after each cut. Once you've cut the first cap, test the fit.

Because the maple stock is wide enough at the center to be trimmed to the diameter of the circle, its ends need to be marked and trimmed before assembly. This must be done evenly or the strap used to clamp the medallion during glue-up may pull the maple caps out of alignment. I use regular carpenter's wood glue to hold the medallion together. It has a relatively short open time, but cleanup is easy. A ratcheting tie-down strap provides even clamping pressure.

TRIMMING THE MEDALLION After the glue has cured (allow 24 hours to be safe), scrape off any squeeze-out and use trammels to draw the circumference around the medallion (see the photos on p. 191). It should just touch the tips of the walnut and cherry wedges. Then set up a router and a circle jig so the bit's cutting edge aligns with the circumference. Use a series of light cuts to trim the medallion round. Install the medallion as explained in chapter 8.

OTHER ROUND MEDALLIONS

The star medallion is constructed using similar techniques to the St. Mark's (see the sidebar on pp. 192–193). The star medallion actually uses the same wedges as the St. Mark's, but with the acute ends pointed

Make It Easy to Duplicate a Medallion

I like to draw out medallions full scale on plywood or cardboard. Plywood and cardboard are rigid and easy to handle, which means you can literally save a hard copy. If you make a medallion once, it's often possible to be called on to duplicate it. When that happens, working from an archived drawing saves considerable time.

Additionally, I like to save one of each of the wedges or pieces of any medallion I build. That makes setting the fences and jigs to duplicate a medallion much easier.

A Quick Math Refresher

The centers of many round medallions are filled with wedge-shaped segments. The first step in making such medallions is determining how many wedge segments you'll need. This depends on the width of the boards you want to use; calculate the circumference of the medallion and divide it by the width of the boards.

Calculating the number of wedges

A medallion with a diameter of 24 in. would have a circumference of 75⅓ in. The circumference of a circle is equal to the diameter x pi, or 3.14. The concept of pi is about 4,000 years old. The ancients probably realized that a wheel rolls 3.14 times its diameter in one revolution. A 1-in.-diameter circle will roll a little more than 3.14 in., so a 24-in. circle would roll 75⅓ in. (24 times 3.14). If we used standard 3¼-in.-wide boards we would need 24 boards (75⅓ in. divided by 3¼ in. equals 24 pieces with just a little wiggle room left over).

Calculating the wedge angle

Calculating the angle of these wedges comes next. A circle has 360°. Divided by 24 segments, we end up with a wedge angle of 15° (360 divided by 24). How do you set a jig to exactly 15°? Protractors are a little too imprecise. Machinists use precision angle blocks to set up equipment, but the method I prefer, and which works for a 1-ft. starburst or a 20-ft. medallion, involves a little high school math.

Any size wedge can be calculated by using its tangent. This is a function of an angle expressed as the ratio of the length of its adjacent side to the length of its opposite side in a right triangle. Obtain the corresponding tangent for any angle using a $5 calculator with that function (or you can look it up on the Internet). Common tangents used for starbursts are shown in the table at left below.

ANGLE (°)	TANGENT
6	0.1051
9	0.1584
15	0.2679
24	0.4452
30	0.5773
36	0.7265

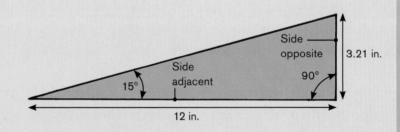

For a medallion whose diameter is 24 in., each wedge will be 12 in. long. That means the adjacent side of the angle is 12 in.

We know the required angle is 15°, and that the tangent of 15° is 0.2679.

Tangent (15) = side opposite ÷ side adjacent
0.2679 = side opposite ÷ 12 in.
0.2679 × 12 in. = side opposite
3.21 in. = side opposite
So, measuring over 12 in. and up 3.21 in. at 90°, then connecting the dots, will produce a 15° angle.

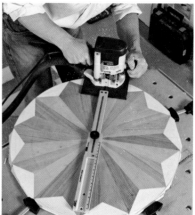

TOP LEFT Verify the center and the diameter of the medallion using trammels.

TOP RIGHT Adjust a circle-cutting jig so the router bit's cutting tip aligns with the circle drawn with the trammel points. Make shallow passes to avoid tearout. Later, a Forstner bit will be used to drill out a spot for epoxying in the brass center disc.

LEFT The St. Mark's medallion installed in the finished floor (next to a starburst medallion).

out. This one was installed along with a segmented curve (see the sidebar on pp. 194–195) and a brass inlay (see chapter 8).

MAKING A STARBURST MEDALLION Another spectacular circular medallion is the starburst. In this design, which is composed of 24, 15° maple wedges, the end of each wedge is cut in a semicircular scallop, which is filled with a matching piece of walnut (see pp. 196–197). The medallion relies on the contrast with the walnut to set off the subtle grain of the maple. You could also make the wedges from contrasting wood. Much of the construction mirrors the previous medallions—cutting wedges from 5⁄16-in. stock and gluing them to a plywood backer. The difference is mainly in scalloping the ends of the wedges and cutting the walnut to fit the scallops.

Making a Star Medallion

Made from $^5/_{16}$-in. veneer with a $^1/_2$-in. plywood backing, this 24-piece medallion is less intricate than the St. Mark's, but it makes an equally impressive inlay. The star is walnut and cherry, while the infill pieces between the star tips are quartersawn white oak. It would be possible to make the star and skip the infill pieces, but that would drastically complicate installation.

The larger arms of the star duplicate the geometry of the St. Mark's medallion segments, but they're installed in reverse orientation. The minor arms duplicate the angles on the shorter pieces.

1. Make all the segments full size, and then mark the smaller star points to length by aligning and overlaying a pair.

2. Since there are only a few, cut the shorter pieces on a miter saw.

3. The white oak infill pieces are wedges cut at 75°. With only a few to cut, the miter saw is again a good way to go.

4. Dry fit all the pieces prior to gluing, and be sure to cut the infills from wide enough stock to allow the medallion to be cut into a circle.

5. Apply moisture-cure adhesive to a piece of ½-in. Baltic birch plywood.

6. Set the segments into the adhesive, being careful not to get adhesive between the joints where it could affect the fit or stain the surface of the unfinished wood. Wiggle the pieces into place to ensure good contact with the adhesive.

7. Set the infill pieces, and secure everything with blue tape. Place plywood on top of the medallion and weight it with a 5-gal. bucket of glue or something similar. Use a piece of paper or plastic beneath the plywood so squeeze-out doesn't glue it to the medallion.

Making a Segmented Circle Border

Round medallions are often bordered by a segmented circle. These segmented inlays are made by gluing wedge-shaped pieces into the shape of a polygon and then cutting the circle, usually with a router guided by a circle-cutting jig.

This circle is ½ in. wide and made from maple. If a circle inlay is much thinner than this, it's usually easier to bend the stock to shape.

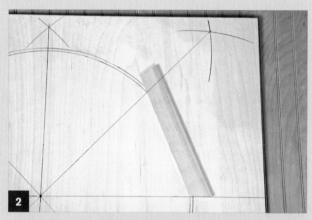

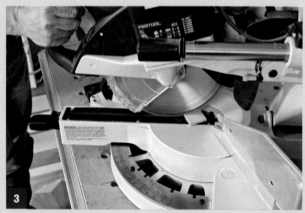

1. Use trammels to draw the inner and outer diameters of the circle inlay on a sheet of ½-in. or ¾-in. plywood. Then lay out the segments with the trammels and a straightedge.

2. This circle has been divided into 8 segments; 360° divided by 8 equals 45° per segment. The miter angle of each segment is half that, or 22½°. Place the maple pieces around the circle and mark the cuts.

3. Set a miter saw at 22½°, and use a stop block to make the repeated cuts.

4. Glue the segments together with 5-minute epoxy, holding them together with small pieces of tape while the glue sets. Place plastic under the work so as not to glue it to the plywood.

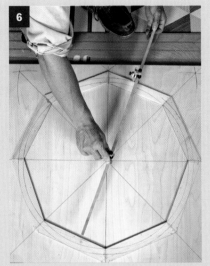

5. Apply hot glue to hold the ring in place temporarily.

6. Once the glue has set, mark the ring outline with trammel points. Brads set outside the cutting area at the outside corners and inner center of the segments bolster the hot-melt glue.

7. Rout the circle using a circle jig and a sharp bit. Make successive passes and a light initial cut to avoid tearout.

8. With the segmented ring glued in place, rout a groove around its perimeter for a brass inlay.

9. Star medallion, segmented circle, and brass ring blend into a cohesive whole.

ABOVE Contrasting arcs of walnut that nestle in the scalloped ends of maple wedges create the look of a craftsman's tour de force, but it's all done with a router and a jig.

TOP RIGHT To make the starburst medallion, begin by cutting 24 wedges of 5/16-in.-thick veneers. Angle the pieces at 15°, and make them a little longer than the radius of the medallion.

MIDDLE RIGHT Fit a router with an inlay bushing, which will follow a plywood template to scallop the maple wedges. The template will be set back from the desired cut by the distance between the bit's cutting edge and the outside of the bushing.

BOTTOM RIGHT Using one template to cut both the maple and the walnut is the key to seamless joints. Draw a centerline down a length of 1/4-in. plywood, and then center a 4-in. hole on this line at one end of the plywood.

Fitting these pieces requires using a custom-made jig and template (see the drawing on p. 198). It also requires the use of a router that's equipped with guide bushings. Guide bushings mount in the center hole of the router base, and a straight bit protrudes through them. The bushing runs along the edge of a template clamped to the workpiece, duplicating the template's shape. Make the cut away from the template by the distance between the bit and the edge of the bushing. This is very useful, because if you use bushings whose radii vary by the diameter of the bit, mating pieces are easy to cut using the same template (see the bottom right drawing on p. 199).

When using a router, particularly while making the delicate cuts required for inlays, a couple of cautions are in order. First, always use a sharp bit. Upcutting spiral bits are a good choice, as they help to clear the chips quickly. However, because spiral bits are made from solid

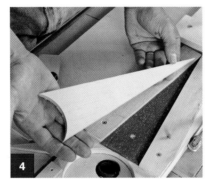

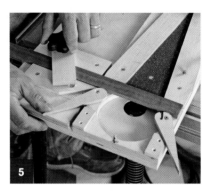

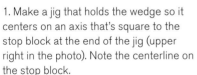

1. Make a jig that holds the wedge so it centers on an axis that's square to the stop block at the end of the jig (upper right in the photo). Note the centerline on the stop block.

2. Make the template so it butts to the stop block and aligns the hole over the maple wedge such that a router guided with a bushing trims the wedge to the proper length. Be certain the template aligns with the jig's centerline.

3. Clamp the template to the jig, and cut the scallop with the bushing-guided router.

4. It's critical that the wedge not move while cutting the scallop. Sandpaper glued to the jig helps hold the wedge in place.

5. Use the cams to secure a strip of walnut in the jig. The walnut must be enough wider than the scallop depth to allow the assembled medallion to be trimmed to a circle whose circumference is just at the tips of the scallops.

6. Change the bushing to one that moves the bit one diameter (of the bit) closer to the template, and cut out the walnut arc. It should exactly fit the scallop in the maple wedge.

7. Join the walnut to the maple with carpenter's glue and blue tape. Make up 24 pieces like this, glue them to a plywood base, and trim the medallion to shape with a router and a circle-cutting jig.

STARBURST JIG

This jig holds the wedge-shaped workpiece and the template (which guides the router) to cut a semicircle in the end of a starburst wedge. The side strips must center the wedge and fit to it tightly so it doesn't move. The template butts to the stop block so it's centered on the wedge. A clamp holds the assembly together.

To cut the contrasting semicircle that fits in the wedges, two cams hold a strip of wood to the jig.

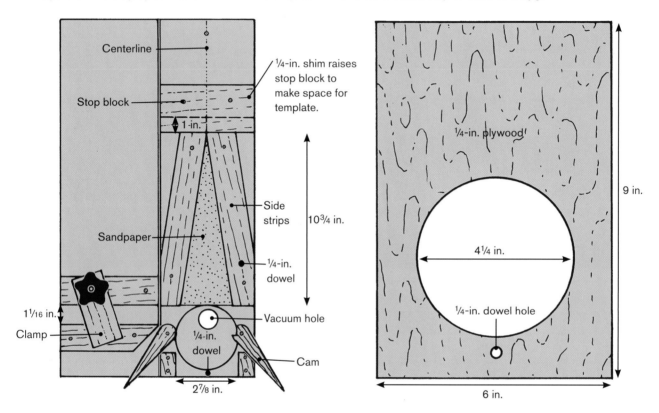

carbide, they tend to be more expensive than bits with straight cutters. They can also be brittle, so go slowly with them to avoid imposing too much lateral stress. This, along with taking light cuts, is good practice anyway in that both minimize tearout. Also, be sure to run the router in the correct direction.

Choose Router Direction Carefully

Routers rotate clockwise when viewed from above. Always move the router in a direction that counters its torque so the router doesn't jump away from you. The router bit teeth are angled to cut into the wood. In Drawing A, the outer perimeter of a medallion is being trimmed to size. The router bit will want to dig in and climb clockwise, so in this case move the router around the medallion in a counterclockwise direction.

In Drawing B, a hole is being cut in the flooring for the medallion. Because both sides of the bit are contacting wood, the torque is equalized and the direction does not matter. In Drawing C, the hole cut in Drawing B had to be enlarged to accept the medallion. The bit would try to climb counterclockwise, so move the router in a clockwise direction.

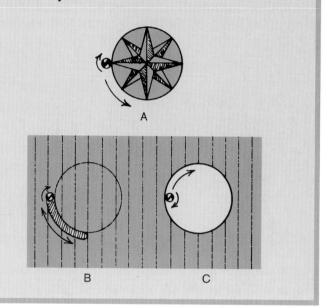

ROUTER GUIDE BUSHINGS

Using router-guide bushings, it's possible to fit inlays together perfectly. The secret is in utilizing two different-diameter guide bushings; the difference in their sizes compensates for the diameter of the router bit.

The smaller bushing (A) is used to cut the piece that will be inlayed. The inside of the router bit cuts the inlay. The template is then placed where the inlay will go. Use a second bushing (B), whose diameter is greater than that of bushing A by twice the width of the bit, to cut out the hole. This larger bushing moves the router bit one diameter away from the template, and the bit's outside cuts out the background piece. When placing the two pieces together, they will mate perfectly.

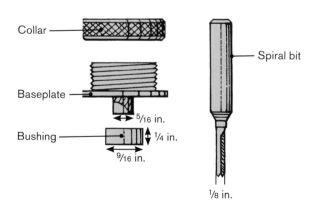

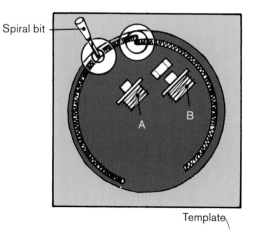

Scroll sawing takes patience and practice but can create magnificent flooring inlays.

HIDE THE GAP

To cut inlays without gaps, slightly tilt the table of the scroll saw. The two pieces will be cut slightly wedge-shaped, with the top surface smaller than the bottom. The trick is to angle the scroll saw table just enough to compensate for the width of the saw blade, which takes a couple of test cuts.

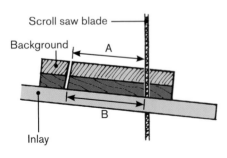

Scroll saw blade

Background

A

Inlay

B

Scrollwork

Scroll sawing can create intricate patterns, from flowers, vines, and leaves to anything you can imagine. If it can be drawn in two dimensions, it can be cut out with a scroll saw. Now, a scroll saw is not a jigsaw. Jigsaws are handheld tools whose blade is relatively thick. A scroll saw is a small bench-mounted saw whose thin blade, held top and bottom, moves up and down. The trick to scroll sawing is cutting out the positive and negative pieces (or the inlay and its hole) at the same time.

I generally use a #2 scroll saw blade that is 0.012 in. thick, 0.029 in. wide, and has 20 teeth per inch. This size blade is generally recommended for cutting veneers up to $\frac{1}{8}$ in. thick. Because the veneers I'm cutting are $\frac{5}{16}$ in., I have to go a little slower. I think it's worthwhile because the thinner blade makes for a tighter fit between the pieces. You might want to start practicing with a #5 blade, which is 0.015 in. thick and will be less likely to distort. A trick that helps to hide the saw kerf is to make the cuts with the saw table tilted.

You can use scrollwork for medallions and borders. I prefer to use $\frac{5}{16}$-in. veneers, which I glue to $\frac{1}{2}$-in. plywood backing. Not only is this more stable, but the actual sawing also goes a lot faster in the thinner material. One other consideration with scrollwork is grain orientation. Think of scrollwork as akin to art painting—it's not just color that makes a painting. The direction of the brush strokes has a more subtle, but important effect. Grain direction works the same way.

Making medallions and inlays is a great counterpoint to the often production-oriented trade of installing floors. Nailing down hundreds of square feet of strip flooring in a day leaves you feeling incredibly productive, if a bit tired. But slowing down to create a medallion that's a work of art not only sets me apart from my competition, but it also provides a different and welcome kind of satisfaction. Another way of saying this is that most flooring can be installed by journeymen while medallions require a master craftsman.

Wood Grain Orientation

When designing a medallion I always start with a drawing. Based on photos of the object I'm copying, I draw arrows to indicate the general direction I want the wood grain to follow. The choice of direction depends on the desired effect. Even slight changes in grain direction will add subtle contrast between the pieces. This effect varies with the species and is more apparent after the finish highlights the grain.

It helps to have a small collection of $5/16$-in. veneers when designing a scroll-sawn inlay. I like to hold the veneer up to a photograph of the object I am copying to decide which wood tones to use and which way to orientate the grain.

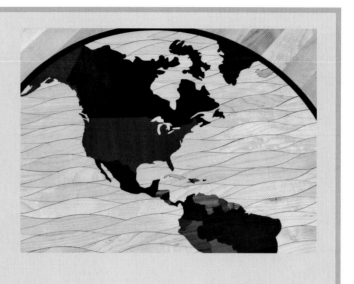

It's relatively simple to cut out intricate mating patterns using a scroll saw. The trick is to cut both at the same time. Tape the two pieces of wood together (here, walnut and maple) and cut along the pattern. There will be a gap between the two pieces that's the width of the blade—in this case, 0.012 in.—but such a narrow gap is acceptable on most floors. (To avoid the gap, angle the table.)

The Victorian grape vine is a classic border design. Because the mating pieces are cut simultaneously, if you go outside the pattern line no one will be able to tell. The two parts will still mate flawlessly.

It is generally not necessary to glue the two pieces together as they both get glued to plywood backing.

CHAPTER TEN

Sanding Wood Floors

FOR A CHANGE OF PACE, I OCCASIONALLY LIKE to make marquetry tables, which entails scroll-sawing and gluing veneers in floral and bird motifs for the tops. It's a nice diversion from working on wood floors. The tables I make are small, but sanding them well takes a fair amount of time. The basic principles of sanding are the same with floors as with tables. Sanding flattens the surface, and a flat surface is critical for finishing. If there are high spots when you sand between coats, it's far too easy to sand through the finish. Beyond flattening, smoothing the surface helps it reflect light evenly and removes scratches that distract from the beauty of the wood.

What to Look for When Renting a Sander

It's always a good idea to inspect rental equipment before leaving the store. You don't want to get to the job and find out there's a problem, and you don't want the rental store to hold you responsible for damage caused by others.

• On a drum sander, check the drum for nicks and dents that could create imperfections in the floor.

• Check the drum's paper-clamp mechanism; you shouldn't have any problems loading the paper tightly.

• On an edger, make sure the rubber-backing disk is flat and the wheels are round and clean.

• Give the power cords a close look because they are often damaged.

• Before leaving the rental shop, ask how to set up and adjust the machines. Wheel and drive-belt adjustments are especially important to a smooth-running sander.

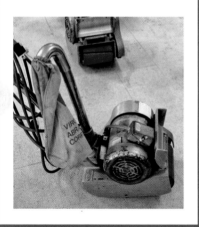

The first job of floor sanding is to flatten the surface. Here, a poor sanding job has created depressions that look terrible in raking light.

Flooring abrasives come in a variety of types and grits, typically bound to tough cloth backings that can take the heat and pressure of aggressive sanding.

One thing that works in your favor with floors is that, because normal viewing is from 5 ft. away, you don't have to end with as fine grit sandpaper as you do with furniture. But unlike a small table, wood floors generally consist of thousands of boards from many trees. Wood floors are probably the largest area that most people would ever have to sand. Efficiently sanding an area that may be a thousand times the size of my tables requires large, powerful equipment.

Floor sanding equipment is available to rent from home centers and rental yards. Usually, rental equipment isn't as heavy duty as the professional gear I use. However, there are ways to use rental equipment, as well as alternative sanding gear, that can still yield a floor ready for a great-looking finish.

Abrasives and Sanding Machines

While there are several types of floor sanding machines, two key points apply to all: First, the direction of sanding, and, second, using progressive grit sizes to obtain the desired smoothness.

Most sanding should be done in the same direction as the wood grain. While cross-grain sanding is faster, and is sometimes used to initially level floors with a lot of overwood (variation in height from board to board), doing so cuts across the width of the cells in the wood, which tears and frays the cell walls. This makes for more finish sanding, and may not result in saving time. More on this later in the chapter.

All abrasives work by scratching away at the surface. Progressing through finer grits simply replaces big scratches with smaller ones. Eventually, the scratches are small enough to be unnoticeable. Coarse grits remove material efficiently but leave a rough surface. Finer grits leave a smoother surface but remove material more slowly. Using progressively finer grit sizes gets you to the desired smoothness most efficiently. The rule of thumb is to start with the least aggressive grit that will get the job done in a reasonable amount of time.

ABRASIVES

Abrasives for wood flooring generally come in a range of grits from 12 to 150 (the terms *grit* and *grade* are interchangeable). The higher the number, the finer the grit and the smoother the resulting sanded surface. The abrasive particles on 12-grit paper are nearly ten times the size of the abrasive particles on 100-grit sandpaper.

There are two common grading systems for flooring abrasives. CAMI (Coated Abrasives Manufacturing Institute) is an organization of mostly American coated-abrasive manufacturers. FEPA (Federation of European Producers of Abrasives) is the European organization that maintains abrasive specifications. Surprisingly, FEPA-graded abrasives are more common in the US than CAMI grades. With the FEPA system, the letter "P" precedes the grit number and designates an abrasive grain used in the sandpaper industry. While there are slight differences in grit size, there's little practical difference between the two grading systems (see the chart on p. 206).

Although the term *sandpaper* is used generically, most flooring abrasives use a cloth backing. I know of no sanding belts that are paper, though the disks used in edgers and multi-disk sanders are frequently paper. Abrasive grains are glued to the backing, and the distribution of the abrasive grains on the backing is referred to as the "coating." There are two levels of abrasive grain density, called open coat and closed coat.

Coarser abrasives are applied to their backing with space between the grains, called open coat. This provides space for the large amounts of material they grind away. Finer abrasives are applied more densely, referred to as closed coat.

CAMI VS. FEPA GRIT SIZE

The two common grading systems used for abrasives in the US use similar numbers,
but the grit sizes vary somewhat.

CAMI			FEPA		
MICRONS	AVG. DIA.	GRIT	GRIT	AVG. DIA.	MICRONS
53.5	0.00209	240	P280	0.00204	52.5
66.0	0.00257	220	P220	0.00254	65.0
78.0	0.00304	180	P180	0.00304	78.0
93.0	0.00363	150	P150	0.00378	97.0
116.0	0.00452	120	P120	0.00495	127.0
141.0	0.0055	100	P100	0.00608	156.0
192.0	0.00749	80	P80	0.00768	197.0
268.0	0.01045	60	P60	0.01014	260.0
351.0	0.0139	50	P50	0.01271	326.0
428.0	0.0169	40	P40	0.01601	412.0
535.0	0.02087	36	P36	0.02044	524.0
638.0	0.02488	30	P30	0.02426	622.0
715.0	0.02789	24	P24	0.02886	740.0
905.0	0.03535	20	P20	0.03838	984.0
1320.0	0.05148	16	P16	0.05164	1324.0
1842.0	0.07174	12	P12	0.06880	1764.0

Open-coat abrasives have space between the grit to make room for the swarf (sanding dust) to be carried until it's picked up by the sanding machine's dust collection. This is more important with coarser abrasives that remove material more aggressively. Indeed, although abrasives for general woodworking are often open coat even in the higher grits, that's

OPEN COAT VS. CLOSED COAT

OPEN COAT

- Mineral grains cover 40 to 70% of backing.
- Used where work surface tends to clog or load coated abrasive surface (e.g., from resinous woods or previously waxed floors).

CLOSED COAT

- Mineral grains cover 100% of the backing.
- Used where loading is less a problem and a finer finish is desired.

not the case with flooring abrasives. Only the extra-coarse abrasives are open coat.

Abrasives are also classified generally as extra coarse (12 to 24), coarse (30 to 40), medium (50 to 60), fine (80 to 100), and extra fine (120 and finer). For the purposes of flooring, 120 grit is about as fine as is useful. I've experimented with up to 150 grit and found that finishes don't always stick as well to wood sanded that finely. And most floors are finish-sanded with 80 or 100 grit.

TYPES OF ABRASIVES Most minerals used to make floor-sanding abrasives are man-made and include silicon carbide, aluminum oxide, zirconia alumina, and ceramic alumina oxide. Garnet, a naturally occurring mineral, is still used sparingly. The perfect abrasive balances the qualities of sharpness, hardness, durability, and friability. Friability is the relative breakdown of the abrasive grain through attrition to expose new, sharp-cutting edges, and it's an important factor in how long sandpaper lasts.

Silicon carbide is the hardest mineral (next to diamond) and is characterized by sharp sliver-like grains that give it a superior ability to penetrate hard finishes. This mineral is typically used in the open-coat abrasives for rough and coarse sanding. Its elongated shape shears off easily, making it too friable for bare-wood sanding because the abrasive wears down too quickly. Silicon carbide abrasive products, which are typically black in color, are cheap, cut quickly initially, and work great at

Grit Size

Ever wonder what the number assigned to abrasive grits means? In the CAMI system, it refers to an ANSI Standard screen size. For example, 24-grit abrasives fall through a 1-in. screen divided into 24. Finer abrasives, such as 120 grit, fall through a 1-in. screen divided into 120.

To provide optimum life, you need to store abrasives correctly. Abrasives perform best if stored at 40 to 50% relative humidity and 72°F. Paper abrasives are hydroscopic—if you store them in a damp place, the paper will absorb moisture and soften.

Sanding Equipment

Several types of equipment are used for sanding floors, ranging from large, walk-behind belt or drum sanders to edgers that get close to walls and into tight areas that the big machines can't reach. The floor buffer or rotary sander is a large walk-behind machine fitted with circular paper used to blend the sanding marks of the big machine and edger or to sand between coats of finish. Smaller sanders are used for areas such as stairs, or for particularly challenging floors. Finally, there's the safety equipment (see chapter 2). I can't overemphasize the importance of protecting eyes, ears, and particularly lungs.

taking off gummy finishes in the lower grits. Since they are inexpensive, they are economical when getting through a lot of belts that are clogging.

Aluminum oxide is extremely hard wearing and has a blocky wedge-shape profile that penetrates tough materials without fractioning, making it well suited for hardwoods. Aluminum oxide is used whenever toughness (the ability to resist fracturing) is the main consideration. One way to think about the differences between silicon carbide and aluminum oxide is to compare them to glass and Plexiglas®, with silicon carbide being glass and aluminum oxide being Plexiglas. Throwing a brick at the glass would shatter it, while the brick would bounce off the Plexiglas. However, I could scratch the Plexiglas with a glass sliver.

Zirconia alumina, also known as AZ, is a blocky sharpened grain that has a microcrystalline structure that makes it self-sharpening in heavy stock removal applications. Typically blue or green in color, AZ products are general purpose, high quality, and available in grades 24 through 150.

Ceramic aluminum oxide is an extruded, dried, crushed, and fired mineral designed for toughness and durability when used under the big floor machine or on high-speed edgers. I have found that the 3M Regalite® abrasive that incorporates Cubitron®, 3M's ceramic aluminum oxide, allows me to cut aggressively for rough sanding with the big machine or edger and in the fine grades cuts velvet smooth; it's my preferred abrasive. Other manufacturers offer similar products.

THE BIG MACHINE

The "big machine" is either a drum or belt sander. Drum sanders employ a sheet of sandpaper wrapped around a horizontal drum. The sandpaper is secured in a diagonal slot on the drum and clamped down. As the name suggests, a belt sander uses a sanding belt instead of a piece of paper wrapped and clamped. These belt sanders aren't like handheld belt sanders though, where a large section of the belt contacts the work below a flat plate. In professional flooring belt sanders, the belt runs around a large drum at the floor level, and around a smaller idler drum above that tensions the belt. Like a drum sander, the sanding belt contacts the floor only at a narrow area that runs the length of the horizontal drum. A second type of belt sander used for rentals has no idler drum. Instead, the belt is placed on a drum that expands to secure the belt.

I prefer belt sanders to drums, mostly because the paper-clamping slot on a drum sander can make chatter marks on the floor. Although professional-level drum sanders are big, powerful machines, those available in most rental yards, as well as belt sanders that use the expandable drum, are relatively lightweight machines that run on 110v. These lack the power and weight of a professional-level sander that runs on 220v, so they take longer to sand the same area.

The big machine, a walk-behind belt sander, does the lion's share of professional floor sanding. Lighter machines are available to rent.

Unfortunately, most home centers and rental yards have only the lighter machines. Because of this fact, and the high cost of buying heavy-duty professional floor sanding machines, occasional users are likely to be stuck using a lighter machine. Although the lighter machines work in much the same way as the heavier ones, they tend to vibrate more and can leave chatter marks on the surface. And where a professional machine might do the job in one pass, it could take two with a lightweight rental machine. The chatter marks from a lightweight drum sander can be smoothed out using a rented buffer equipped with a thin pad and sanding screen. Another option, though slower, is to rent a multi-disk sander. More on this later.

Professional big machines are available in widths of 10 in. or 12 in. The rental sanders tend to be narrower—typically 8 in. All have integrated dust-collection systems, generally rated from 200 cfm to over 300 cfm, which exhaust into a standard 5-micron cloth filter bag. Units with ratings of 300 cfm have a tendency to induce chatter marks into the flooring caused by vibration from the fan. I prefer machines rated closer to 200 cfm.

If you were to start a big machine with its drum contacting the floor, it would make a mess of that spot. Modern big machines use a clutchlike device that allows the operator to raise or lower the sandpaper to the floor, which is always done when the sander is moving. Raising and lowering the drum gradually helps to feather in the start and end of each pass, avoiding abrupt transitions. Old-time sanders lacked the clutch and rocked back and forth to raise or lower the abrasive. These machines are rarely found today, for good reason.

EDGERS

Edgers are powerful handheld circular sanding machines able to reach the areas inaccessible to the big machine. I always use the edger against the wall of the room, for example. They generally consist of a 7-in. shrouded disk with a dust collection fan rated at about 110 cfm. The sanding dust is generally exhausted into a 5-micron filter bag. The two wheels on the housing of the edger hold most of the machine's weight and are adjustable to vary the depth and angle of cut.

A variation of the edger, commonly called a "duckbill," can reach under tight areas such as heating radiators and cabinet toe kicks. The

Edgers are aggressive handheld sanders that get into tight places. The elevation of the wheels determines the depth of cut.

A "duckbill" edger is used to sand below radiators and cabinet fronts.

duckbill edger's sanding pad is offset from the main body and driven by a pulley and belt arrangement.

In addition to the edger, I use a handheld orbital sander such as you might find in a cabinet shop on every job. Mine is a Festool 150 FEQ (www.festoolusa.com). Not as aggressive as an edger, it still has incredible power and dust extraction. I have actually sanded entire floors with it. While that's not a regular practice, I sometimes encounter 200-year-old plank floors that are so wavy that sanding with a walk-behind machine of any sort would be impossible. The handheld orbital sander allows me to follow the original contour of the floor. To avoid leaving machine marks on an antique floor, I generally finish with a light hand scraping and hand sanding of the entire floor. Scraping removes any remaining scratches quickly, and hand sanding leaves a uniform surface for finishing. I mainly use the orbital sander for the final sanding around the perimeter, and on stairs.

More commonly found in cabinet shops than with flooring crews, handheld rotary sanders combine a fair degree of aggressiveness with great controllability and superior dust collection.

ROTARY SANDERS

The rotary sander or floor buffer is a large walk-behind machine generally available with circular driving pads with diameters of 13 in. to more than 22 in. The machine I own is a 16-in. model. Smaller machines tend to be

Dust Collection

Most sanding equipment has its own integral dust-collection system, typically using a fan that runs off the main motor to collect dust and transfer it into attached cloth filter bags. All airflow is directed through the filter bags.

The filter bag needs to be emptied often to maintain proper airflow. Manufacturers generally provide a mark on the bag indicating when to empty (generally the halfway mark). The bag itself is the filtering medium. The fuller the bag becomes, the less filter area is available for airflow thus minimizing dust pickup. The filter bag should be maintained by turning it inside out and vacuuming to remove fine dust particles that clog the filter pores. Some manufacturers also have bags that can be washed.

Rotary sanders or buffers have several uses. First, they're used after the main sanding is done to blend the sanding marks. Second, they're used to sand between coats of finish. The black disc is a sanding screen. The white one is a synthetic driving pad that links the screen to the sander.

lighter and less aggressive. Buffers are available to rent. The buffer's pad can be fitted with circular sanding paper, abrasive screens, or abrasive pads. I generally use the buffer to blend the sanding marks from the big machine and the edger. The different sanding motions of the two machines can leave a noticeable band around the room (known as "picture framing") if a floor is being stained.

Buffers run at low speeds of approximately 175 rpm, but you still have to be careful as buffers can cause dishing, particularly if used with a thick pad below the screen. Thick pads allow the screen to move up and down, which scoops out the softer wood and leaves the harder wood. To avoid this, use thinner pads, usually around ¼ in. (pads range up to about 1 in.). Newer buffers are available with integrated dust-collection systems. Older units are adaptable by attaching a dust skirt and boring a hole in the body to attach a vacuum system.

MULTI-DISK ROTARY SANDERS

Multi-disk sanding machines are sometimes used to sand wood flooring and can be rented in most areas. Some walk-behind units have four 6-in.

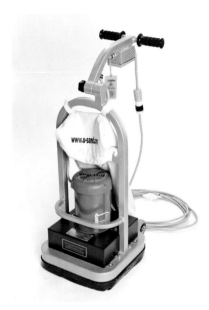

orbital sander heads. Other machines, called tri-planetary sanders, have three 8-in. orbital heads spinning in one direction attached to a main disk that spins in the opposite direction. Either of these machines is much better for ornate floors than using a combination of the big machine, edger, and buffer. Multi-disk machines do not have to be oriented to the direction of the wood grain and don't tend to dish out softer wood cells. They are not as aggressive as the big machine or edger, but they're easier to do a good job with.

EXTERNAL DUST COLLECTION SYSTEMS

Even though most floor sanding equipment has integral dust collection (see the sidebar on the facing page), a high-quality vacuum system is a virtual necessity for cleaning the floor of dust and abrasive grit between sandings and before finishing. The sanders' dust collection isn't 100% efficient, and most of the bags filter down to only 5 microns, which means a fair amount of dust may not be completely filtered out. Even though machines have dust collection, it's still important to wear a 1-micron respirator. This is because the particles in the 1 to 5 micron range are the most dangerous ones to get in your lungs. Larger particles can be coughed up, and smaller ones are picked up by the blood and removed from the lungs. Several manufacturers have vacuums that tools

Toxic Wood Dust

Wood flooring professionals face frequent and long-term exposure to wood dust. Wood dust generated by floor sanding is considered a toxin and carcinogen. Although the level of toxicity varies considerably according to the species of wood, you should make all attempts to avoid direct skin contact and breathing the dust. Flooring installers who smoke or have respiratory or sinus conditions are at an even higher risk for health problems.

Wood dust is either an irritant or a sensitizer. Irritants cause a fairly rapid reaction. Sensitizers may have a latency period of hours or even months and may require repeated handling before any reaction occurs. Once you are sensitized, the reactions only get more dramatic with each exposure. Although uncommon, certain sensitive individuals can experience anaphylactic shock, a life-threatening condition where the airway swells shut. Additionally, some woods are known carcinogens.

KEY
I = irritant
S = sensitizer
C = cancer causing
P = hypersensitivity, pneumonia
DT = direct toxin

WOOD DUST TOXICITY

WOOD TYPE	REACTION	POTENCY LEVEL
Olivewood	I,S	High
Beech	S, C	Medium
Padauk	S	Low
Birch	S	Medium
Elm	I	Low
Purpleheart	S	Medium
Iroko	I, S, P	High
Redwood	S, P, C	Medium
Rosewoods	I, S	Very High
Satinwood	I	High
Sassafras	S, DT, C	Low
Mahogany	S, P	Low
Maple	S, P	High
Spruce	S	Low
Walnut, black	S	Medium
Wenge	S	Medium
Oak	S, C	Medium
Western Red Cedar	S	High
Teak	S, P	Medium
Yew	I, DT	Very High
Zebrawood	S	Medium

(Adapted from the 10th Report on Carcinogens, U.S. Department of Health and Human Services.)

such as sanders plug into, and which start up when the tool is switched on. One-micron bags come standard with these machines, which mean they exhaust very little dust. Unfortunately, at this time, 1-micron bags are not available for floor sanders because they restrict the airflow too much and wouldn't be effective.

Additionally, several manufacturers sell separate dust containment systems that are mounted in a truck or trailer and hook up to the sanding equipment with long hoses. One advantage is that any fine dust that escapes the filters does so outside the house. The systems need to be very powerful to compensate for the restriction losses from the hoses. These systems are extremely expensive.

Yet another approach is a dust collector on wheels, such as is commonly used in small wood shops. One-micron filters are available for these machines, which can cut down on dust. However, like the remote dust collection systems, you're stuck dragging around a hose. And, because these machines don't have enough power required to overcome the friction losses imposed by long hoses, they have to run fairly close to the dust source.

Offering better filtering and a larger bag capacity, specialized dust extraction vacuums can handle smaller sanders such as an edger or handheld rotary sander. They're also ideal for vacuuming between grits.

Sanding New Floors

Newly installed flooring is generally the easiest to sand. The biggest problem is that the butt ends of individual boards often aren't flush and take a lot of sanding to level initially. On the positive side, there are no old finishes to gum up the abrasive. The first step in finishing wood flooring is to sweep or vacuum the floor thoroughly and inspect for protruding fasteners. It takes just one protruding nail head to damage the sanding equipment. All nails should be countersunk deeply enough so that they don't cause damage or produce sparks when hit with the abrasive that can cause fires. A spark sucked into the sanding bag is force-fed air and fuel, and fires have started this way.

Even if the floor was installed at optimal moisture conditions, check it again before sanding and finishing. The wood flooring should be acclimated to a moisture content that corresponds to normal living conditions prior to sanding, since wood changes dimensionally with moisture (see chapter 1). If the floor was glued down, make sure enough

Before sanding, find and set any protruding fasteners. They'll tear up the abrasives, damage drums and sanding pads, and if they spark, can create a fire in the super-charged atmosphere of a dust collection bag.

Wood flooring is rarely perfect, and variations in its manufacture or installation result in overwood, where the surfaces of adjacent boards are not flush. It's a common condition with freshly installed floors, worsened by poor installation or low-quality flooring.

time was allowed for the adhesive to cure. Some installers allow mechanically fastened wood flooring a few days to allow internal stresses to equalize.

Sanding wood flooring first means removing all the factory milling marks and "overwood." Overwood is the difference in height from board to board. The degree of smoothness required of a wood floor is determined by the finish being used. This will vary with the finish manufacturer, and the instructions on the finish can are usually quite explicit. A typical sanding sequence might be 50 grit then 80 grit with the big machine, followed by screening with a buffer at 100 to 120 grit.

The overwood in the floor shown here wasn't bad, but boards in a poorly installed or poorly manufactured floor might vary in height by the thickness of a nickel. When floors are that bad, I'll often grind down the worst of the overwood with an edger, feathering out over several square feet, before beginning to sand with the big machine. Here, I could feel the overwood with my fingers, which was good enough to allow me to start at 60-grit paper cutting at a 20° angle and 80 lb. of drum pressure on the big sanding machine.

Sanding a floor takes a sequence of grits. Each successive grit should erase scratches left by the previous grit. Most flooring professionals, myself included, sand with every other grit in the sequence. The final grit is determined by the finish manufacturer's recommendations. The grit sequence is determined by working backward from the finest grit recommended for the finish being applied. For example, if the final grit recommended by the finish manufacturer is 120, I would start with 50, skip 60, sand with 80, skip 100, and finish with 120.

Some species of wood have particular sanding and finishing nuances. Maple and hickory are good examples. Because they are very hard, and maple is close-grained as well, scratches show up and are difficult to remove. I'm careful not to sand maple with extremely coarse grits, preferring to start with 60. I don't recommend using extremely coarse grits such as 36 or 40 on it at all. Finer abrasive grits have a tendency to

SKIPPING EVERY OTHER GRIT

When using a big machine or an edger, following a sequence such as shown here of coarse (36 grit), medium (50 grit), and fine (80 grit) will sand floors efficiently and with good-looking results. You can add the intermediate grits of 40 and 60 but that adds time to the job without a similar increase in quality. On the other hand, skipping the medium grit may remove only the tops of the deeper scratches left by the coarse abrasive, and will lead to an unsatisfactory surface.

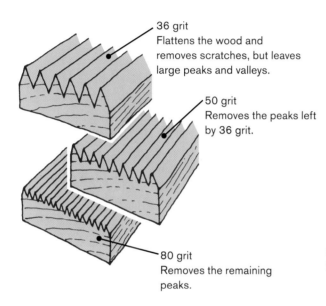

36 grit
Flattens the wood and removes scratches, but leaves large peaks and valleys.

50 grit
Removes the peaks left by 36 grit.

80 grit
Removes the remaining peaks.

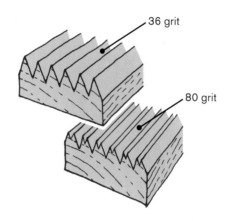

36 grit

80 grit

Skipping all the way to 80 grit removes only the tops of the peaks left by the 36 grit.

RECOMMENDED GRIT GUIDE

TYPE	GRIT	PURPOSE
Open coat	12, 16, 20, 24	Remove old shellac, varnish, or paints, starting with a diagonal cut followed by sanding with the grain
Coarse	30, 40	Fast leveling of uneven floors, starting with a diagonal cut followed by sanding with the grain
Medium	50, 60	First sanding of a new floor
Fine	80, 100	Final sanding

Setting Up a Big Machine

Sanding machines all require a few steps in preparation for use. Most should be undertaken with the machine unplugged for safety. Use care when lifting the machines as they're extremely heavy, and be careful working around the bottom of the machine, as years of use often leave sharp edges.

1. High-quality drive belts transfer the electric motor's power to the sanding drum and the fan. Because they cause vibration, any imperfections on the belts will impart imperfection to the floor. Any time the sander stops for more than 10 minutes the belt should be de-tensioned so as not to create imperfections in the belt.

2. Always use the outer sanding belt in a group first because it is deformed the least from storage. Using the inner belt last gives it time to relax into its proper shape. Do not use any force to pull the belt out or you may damage the backing and remove abrasives. Just flex the inner belts and the outer one will fall off.

3. Rotating the drum in a clockwise direction can help when installing the belt. Rotating the belt counterclockwise will aid in removing it. Some belts can run in both directions, others only in one. If the belt is omni-directional, there will be arrows inside the belt indicating the proper orientation.

4. Always start the sander with the door open to observe if the belt is tracking correctly and not rubbing. If it is rubbing, the machine can be damaged and the belt may eat itself. An adjustment screw is located on the upper roller, but each machine differs. Consult the directions.

5. Keep the wheels of the sander clean, but take care not to damage them while cleaning. Here, a floor scraper is barely touching the wheel, just enough to knock off any surface particles. Any imperfections on the wheels will telegraph to the sanded floor.

6. Adjust the drum pressure before starting to sand: higher pressures for rough sanding, lower pressures for finish sanding. Many machines come with three factory settings, while others adjust through a range.

Make the first pass with the big machine, using a coarse grit at a slight angle to the flooring to level overwood. Set the sander to its heaviest pressure.

For the second pass, use the same coarse grit as before and sand the floor parallel to the boards.

dull quickly, making it difficult to remove scratches and easy to burnish the floor, closing off the wood pores.

THE FIRST PASS

I always start with the least aggressive grit that will level height discrepancies between floorboards. Usually that's 50 or 60 grit, and I'll make the first pass over the floor at a 15 or 20° angle, followed by sanding in line with the grain using the same grit. Sanding pressure and technique may vary for each installer. In theory, the most efficient progression would be to sand using every grit, if not for the time associated with actually changing abrasive paper. In practice, the most efficient method of sanding ends up being removing all the scratches left by each grit, never skipping more than one grit size.

On the first pass, the greater the angle, the more aggressive the cut, which can be helpful when there's a lot of overwood or when flattening an old floor or removing old coatings. On the other hand, the greater the angle, the harder it will be to remove these angled scratches in subsequent passes. There are a couple of good reasons to start sanding at a slight angle. First, there is typically overwood at butt joints in the

When a floor has significant overwood, tape a pencil to a stick and draw a series of lines across the boards. You'll know you've gotten all the overwood when the pencil lines disappear.

SANDING A ROOM: THE BIG PICTURE

Most of the time, the first sanding of a room should be done at an angle of 7 to 25° to the length of the boards, followed by sanding with the same grit parallel to the floorboards. Sand two-thirds of the room going in one direction, overlapping the passes by 2 to 3 in. and staggering where you raise the sander every couple of passes by a couple feet. When you finish that side of the room,

reverse the sander and repeat the process for the remaining third of the room. Over lap the stop marks of the last pass by at least 3 ft. When you change grits, move the area where you change directions to minimize the overlap. Avoid placing this area directly in front of a light source such as a window.

Sanded direction

7 to 25°

Sander stops
staggered 2 ft.
to 3 ft.

flooring. These act like speed bumps. As the sander bounces over the bump, it takes a divot from the floor. Starting the machine's wheels at a slight angle over the flooring's proud butt ends limits telegraphed imperfections. Any imperfections will be at an angle easily taken out by succeeding straight sanding. Also, the butts are end grain, which requires more aggressive sanding. Overlap the passes by 2 to 3 in.

The second reason for cutting at a slight angle is that it allows a finer grit paper to be used initially. At 25°, I might be able to use 60 instead of 40. Sandpaper cuts more aggressively when sanding cross grain. The entire floor should then be sanded straight with the same grit size to remove all the cross-grain scratches. When approaching within a couple feet of the wall at the end of each pass, I gradually lift the drum from the floor. This feathering technique avoids creating a sharp transition that would require excessive blending with the edger.

Sanding with the Big Machine

As a rule, belt and drum sanding machines should move across the room's width from left to right. The key is that the sander's wheels should always ride on the freshly flattened floor behind the sanding drum. The elevation of the wheels relative to the drum is what determines how much wood to sand off. If either wheel is on the unsanded floor, any irregularities replicate in the sanding.

One pass means both walking the sander forward and dragging it backward over the same section of floor. As you begin to move, gradually lower the sander, and as you approach either end of the pass, gradually raise the sander. This feathers in the start and stop points. Move over so that the next pass overlaps the last by 2 to 3 in. Walk at a steady speed to ensure even sanding. Also, be sure the wheels are clean and smooth. To practice properly raising and lowering the drum start by using the finest grit belt available and set the drum pressure to the lowest setting. Practice in the least conspicuous area of the room.

All the wheels of a sander should ride on a freshly sanded surface because the imperfections in an unsanded floor will telegraph through the wheels to the drum to the sanding surface. Because of the orientation of its wheels, successive passes with this sander should always move from left to right.

As a visual aid for maintaining the proper overlap between passes, apply a piece of 2-in.-wide tape on the sander above the edge of the belt.

I usually sand about two-thirds of the floor in one direction, and then turn around to sand the remaining third. It's important to feather in where these two areas meet or you'll create a noticeable transition line. Feathering is accomplished by slowly lifting the drum off the floor over a 2- to 3-ft. area. To prevent making a discernable low area across the width of the room in the overlap area, randomly stagger the ends of every two or three passes by a couple feet. Make sure the overlap doesn't align with a light source such as a window. Move the overlap with every grit sequence

Because it's so fibrous, do not sand bamboo flooring at an angle. If you do, it will fuzz up and be very difficult to sand smooth. Always make sure to sand bamboo with the grain.

Modern big machines use a clutchlike device to raise or lower the sandpaper to the floor. Begin at a slow walk and gradually lower the drum. Whenever the drum is touching the floor, the machine must be moving. Raise the drum prior to stopping the sander or it makes a deep drum mark (right photo).

so as not to create a noticeable trench across the room. Remember, a wood floor does not actually have to be flat, but it has to look flat.

Your walking speed should be consistent with each grit, though different speeds are called for with different grits. Walk slowly with coarser grits, making sure to flatten the floor. Walk a little faster with medium grits, and fastest with fine grits. You'll be removing less material with each pass, so you should spend less time sanding. Also, heat buildup is a concern with finer grits if you move too slowly, and burnishing the floor is a possibility. Heat buildup is less of a problem with coarser grits because the large amount of swarf removed takes a lot of the heat generated by the friction with it. Finer abrasives remove less material, and so the heat builds up more quickly. Drum pressure also affects heat buildup, so use the heaviest pressure with the coarsest abrasive, medium pressure with medium abrasive, and the lightest pressure with fine abrasive.

Belt and drum sanders have an adjustment to change the amount of drum pressure (i.e., how hard the drum will push the sanding belt to the floor). Professional-grade sanders usually provided set points from 50 to 85 lb. Some machines can provide infinite adjustment through a range. A 50-lb. setting might be used for soft pine with pitch, whereas an 85-lb. setting might be for leveling floors in rough shape. Higher pressures are used with rougher abrasives. Lower pressures are used on finer grits to limit heat buildup and burnishing. Because rental machines are rarely

Sand the floor in line with the boards using a medium-grit abrasive, followed by edging and vacuuming, repeating all steps with the finest grit recommended by the finish manufacturer.

Abrasive grains come off the paper during sanding. Larger grains from earlier grits can gouge the flooring during successive sanding. Also, dust hinders the performance of the next abrasive. Vacuuming before moving to finer grits prevents both problems.

as heavy duty as professional machines, the settings on them are lighter. As a result, whereas a pro machine might do the job in one pass, a rental machine could take two.

In most cases, it takes only one pass with each grit. Particularly with the initial sanding, a lot of material is removed. Keep an eye on the sander's dust bag, and empty it when it reaches the full line, or when it's about half full. The exception is in the initial sanding, which levels the overwood. It's time to change the abrasive on a big machine when you feel a significant change in how it pulls you across the floor. When sanding an old floor, you get another clue—the color change as you sand down to new wood lessens as the abrasive wears out. When using an edger, you'll see when it's no longer cutting.

GOING THROUGH THE GRITS

After the initial leveling, the flooring must be sanded again with the same grit in a straight direction. It is a common mistake for craftspeople to jump to the next size belt and try to remove the cross-grain scratches left by the rougher grit. This can leave cross-grain scratches. Do not do it.

The step after finishing sanding with any grit is always sweeping or vacuuming the floor. Vacuuming does a better job. Before the final sanding with the big machine is the time to fill any nail holes in the

Fill nail holes before the final sanding, and be sure to allow the filler sufficient time to dry.

floor. There are commercial fillers available from flooring suppliers, but be sure they're compatible with the stain or finish that will be used. I prefer to make my own filler with fine sawdust (at least 100-grit edger dust) and Wood Flour Cement (Glitsa®, www.glitsa.com). Not only does this dust-based filler tend to match the wood color better than commercial fillers after staining and finishing, but it also changes color with exposure to UV light more like the surrounding wood. Depending on how much the filler shrinks, a second coat might be needed.

The final sanding is done all in line with the flooring, using the abrasive grit recommended by the finish manufacturer and sanding as close to the walls as you can come without damaging them. On a new wood floor, I determine the final sanding grit based on the finish I will apply. I generally final-sand floors that will not be stained and which will receive a water-based finish with a 120-grit belt. This would mean the grit before the 120 belt would be 80 grit. My first grit would be 50-grit paper cutting at a 15 to 20° angle and 80 lb. of drum pressure on the big sanding machine. If this grit did not work, I would have to drop down to a rougher grit belt.

EDGING AND DETAILING

Around the perimeter of the room, there will be a band that the big machine cannot reach. The edger is used to sand this area. Edge between grits when you're using the big sander, using the finest grit that will flatten the floor in a reasonable time. Often, this will be a finer starting grit than for the big machine. On this floor, I tested an 80-grit disk, but felt it was not cutting efficiently so I dropped to a 60-grit disk. The final grit will be the same as with the big machine.

Before I plug in the edger, I inspect it to make sure the pad is in good condition (no tears or divots) and the wheels roll freely. Never plug in an edger unless it is lying on its side. Most edgers have a toggle switch, and if you get in the habit of plugging them in while they're resting on the pad, it's only a matter of time before you see it spin across the floor, merrily cutting divots as it goes.

Many flooring professionals use different techniques, from moving the edger in a "J" pattern to circular motions. I just run it in a straight line, which keeps the cutting portion of the pad cutting parallel with

MAKING AN EDGER CUT WITH THE GRAIN

Only a quarter-size area at the front of an edger's abrasive disk actually contacts the floor. It's possible to take advantage of this to keep the sanding mostly in line with the grain, even when sanding where the butt ends of strip flooring meet a wall. Minimize cross-grain sanding by turn-ing the sander so the cutting spot on the disk is as far to the right as possible. Tight to the wall, the sander's shroud will limit this, but as you move farther out, turn the sander to eliminate cross-grain sanding.

SLIGHT CROSS-GRAIN SANDING

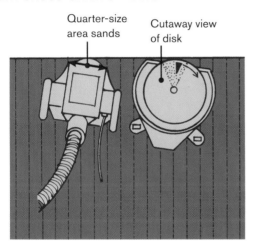

Quarter-size area sands

Cutaway view of disk

PARALLEL SANDING

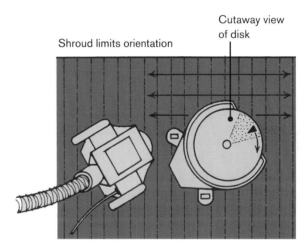

Shroud limits orientation

Cutaway view of disk

Running an edger straight along the wall makes quick work of flattening the areas that the big machine can't reach. To save wear and tear on his back, the author braces his elbows against his knees while edging.

ABOVE You can substitute a handheld rotary sander for the edger for the last sanding with 80- or 100-grit disks. It requires less strength than an edger, and does a great job feathering in the sanding marks.

TOP CENTER Hand scrape the corners where the machines can't go. Pulling the scraper at a slight angle to the grain direction helps to prevent scooping out soft springwood.

BOTTOM CENTER After scraping, hand sand corners with the last grit used by the big sander.

RIGHT Detail sanders get into corners faster than hand sanding.

the grain. The area that the edger pad actually cuts is only about the size of a quarter. Many manufacturers set their edger to cut at 12 o'clock.

Edging is hard on the back. To make it a little less taxing, I rest my elbows on my knees while edging. Many times, I will mark the overwood along the edge with a pencil to help ensure I fully flatten that area. When working right up to the wall, the edger has to be held at 12 o'clock so the shroud allows the disk nearly to contact the wall. As I move the edger away from the wall, I am able to turn it clockwise to align the abrasive scratches with the wood grain. Many times, I'm able to use a handheld orbital sander for the final edging.

No matter whether you're using an edger or a handheld orbital sander, round disks can't reach all the way into corners. Here, I use a variety of tools. The flooring craftsman's standard is the hand scraper. Back in the day, entire floors were scraped flat. If you think using a sander is a tough day's work, try scraping a couple hundred square feet of oak. Still, today professionals consider it one of their most important tools. Flooring scrapers have 1- to 2-in. blades, which can be resharpened.

Scrapers leave a very smooth finish, and I follow up with hand sanding to blend it in with the rest of the floor. Even faster is to use a detail sander. In either case, use the same grit as on the final pass with the big machine.

Floor Scrapers

Scrapers are an essential tool for flooring work. They get into places that sanders can't, such as under radiators, corners, and around the balusters on stairs. Using them may seem simple, but many craftspeople consider it almost an art.

Pulled parallel with or at a slight angle to the grain, a sharp scraper can be very aggressive or cut the finest shaving imaginable. The angle of the handle to the floor is one way to control how aggressive the scraper is.

Raise the handle to take a deeper cut, lower it for finer cuts. Until you've used a scraper, it's hard to appreciate how well they work. That is, when they're sharp.

To many people, sharpening is mysterious and a little bit daunting. At its simplest, sharpening a scraper entails maintaining a bevel angle of about 45° with a slight hook or burr at its edge to do the cutting. The angle of the bevel and hook also affect the aggressiveness of the scraper, and can be adjusted.

BUFFING IS THE FINAL STEP

Equipped with a screen or sandpaper that's typically one grit finer than the final one used with the big machine, a buffer can be used to remove and blend scratch patterns from the big machine and edger. The buffer motor rotates counterclockwise at approximately 175 rpm. The most aggressive cutting area under the buffer is located between the 2 o'clock and 3:30 area with the handle considered 6 o'clock.

Traditionally, buffers drove a sanding screen that attached to the machine by way of a nonwoven fiber pad However, 3M (www.3m.com) has introduced sandpaper disks with hook and loop backs that attach to a thin white buffer driver pad. The sanding disk provides superior results, with less of a tendency to dish out soft wood and leave scratches than do sanding screens.

It's important to use the buffer in the correct pattern in order to prevent cross-grain scratches. Overbuffing the floor with screens with a thick pad will dish out soft grain. Failing to vacuum abrasive grit off the floor will cause scratches. If the room is big enough where the abrasive needs to be changed, don't change it in the center of the room. The sanding should start from one wall and work toward the

Use a buffer to blend in the scratches from the edger and the big machine.

If the handle is at 6 o'clock, then
the only part of most buffers that is
in contact with the floor is the area
between 2 o'clock and 3:30. Keep
this area perpendicular to the wood
grain as you move so its motion is
parallel to the grain. As you complete
each pass, turn the buffer 180° and
go back down the same path. Since
only one side of the buffer cuts
aggressively, this will ensure a uniformly abraded floor.

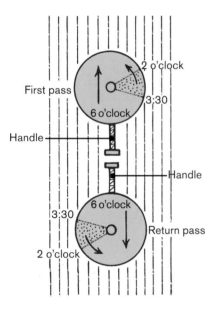

center of the room. Then, change the abrasive and start sanding from
the opposite side of the room toward the center. This avoids there being
a noticeable difference in the center of the room where the dull worn
abrasive was changed to a more aggressive new abrasive.

Once you're done with the buffing, vacuum the room again and
then you're ready for the final step: finishing, which is the subject of
the next chapter.

SANDING RESINOUS WOODS

Sanding resinous woods such as heart pine, southern yellow pine, and
fir can be challenging because sap or resin quickly clogs most abrasives.
Start with 36- or 40-grit paper on a diagonal of about 30°. In some cases,
you may have to drop down to an open-coat sandpaper as low as 12 grit.
Remember always to follow by straight sanding with the same grit. Next
sand the flooring with the grain using the next grit in the sequence. The
paper may require frequent changing.

With an edger, start with 36-grit paper. As with waxed floors, I
prefer to use my Festool 150 EQ in the aggressive mode with their
40-grit Crystal paper. The abrasive stays cooler and cleaner. Remember
to vacuum or sweep the floor frequently. If using an edger, follow up
with 50- or 60-grit paper. The edger wheels have a tendency to crush the
soft wood fibers and leave wheel marks. The last pass should be with the
big machine to remove the edger wheel marks. While this might seem
out of sequence, within the same grit it doesn't matter if you use the
edger or the big machine first. It's a bit like mowing and weed-whacking
a lawn. You can weed-whack first and mow second, as long as you weed-
whack out to where the mower can go. If need be, you can further blend
the edges using a handheld orbital sander. The last pass with the big
machine might only be able to go as high as 50 or 60 grit. Finer grits clog
with resin, which can ball up and actually scratch your newly sanded
floor. Make sure to feather in and remove marks from the edger wheels.
Be careful not to burnish the floor while working with grits 50 and higher.
Screen the floor with 80- or 100-grit screen on a buffer. The screens will
also load up and scratch the floor, so clean and change them often.

BUFF IN A PATTERN

Buff the outer perimeter first using an egg shaped motion. This motion ensures the area is throughly covered. After sanding the entire perimeter, start sanding with the direction of the flooring from one wall working toward the center of the room. If the abrasive needs changing out halfway through, start with the opposite wall and finish in the center. Never change abrasive pads near a light source such as a window or the difference in the scratch pattern may show.

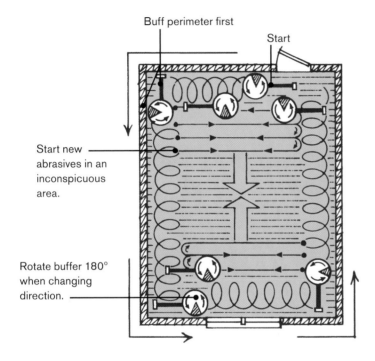

Buff perimeter first

Start

Start new abrasives in an inconspicuous area.

Rotate buffer 180° when changing direction.

SANDING WITH MULTI-DISK SANDERS

The best sander I've come across for DIY use is the four-head random orbit machine made by Cherryhill Manufacturing, the U-Sand (www.u-sand.com). It's widely available in rental stores and home centers. While it's slower than a professional-level big machine, those are hard to rent. And it's not much, if any, slower than a typical rental drum or belt sander if you pay attention to a few details. Another advantage the U-Sand offers is that, because of its straight-sided design, you can do it all with one machine. If you were to rent a drum sander, you'd also need to rent an edger and a buffer.

Sanding Ornamental Flooring

Ornate floors are generally composed of boards of multiple species with different hardness, and they are generally installed in patterns with varying grain directions, which makes it virtually impossible to sand with the grain. There is one cardinal rule to sanding ornate flooring: Never sand perpendicular to the grain. The best way to go is to use a multi-disk sander like a U-Sand or a tri-planetary sander and hand sander like the Festool 150 EQ. I do, however, use the big machine on straight aprons, and finish the log-cabin corners with a handheld orbital sander.

Because it's an orbital sander, direction doesn't matter with a U-Sand, which makes it a great choice for sanding ornate floors. In fact, this is my machine of choice for all the ornamental floors shown in this book. A final advantage is this: Big machines have a learning curve, and sometimes even pros screw up and damage floors with them. While you can screw up a floor with a U-Sand, you'd really have to be trying.

Because the U-Sand isn't as aggressive as some sanders, it's recommended to start out using 50-grit paper. If there is a lot of overwood, you may want to use 40 grit or even 36 grit, but try the finer grits first. Once the floor is leveled, that is, all the overwood is sanded flush, proceed through all the grits—36, 40, 50, 60, 80, and 100. Don't skip any grits and always vacuum between grits. I've found that if you skip a grit, it's very difficult to remove the scratches from the coarser abrasive. Even though the U-Sand has an effective dust collection system, vacuuming is a critical step. Abrasive particles from the previous grit left on the floor can get trapped under a sanding pad and leave a swirl-pattern of scratches. The only fix when that happens is to re-sand the damaged area with a coarser grit.

The U-Sand won't get all the way into corners, but neither will any sander. You'll still need to scrape and hand-sand corners.

Sanding a Previously Finished Floor

Floors wear out in time, and eventually there won't be enough wood left to sand again. The limit for ¾-in. solid flooring is when the wood above the groove is near 3/32-in. after sanding. When the top lip becomes too thin, it will crack and splinter. If the flooring boards are at all loose and move when walked on, all bets are off.

You can check to see how much flooring is left by removing a heating vent or transition. Another method is to insert a business card between the flooring boards until it bottoms on the tongue and mark the depth with a pencil. It's a good idea to check in several spots around the room or rooms. If the floor isn't flat, or it's cupped, and the wear layer is marginal, it may not be possible to sand it again. And some engineered floors come with a wear layer that is too thin to sand.

Refinishing an existing floor requires not only moving furniture but also objects such as radiator covers that would interfere with sanding or be at risk for damage. (This also allows the radiators to be vacuumed clean of years of debris that could otherwise fall into the new finish.)

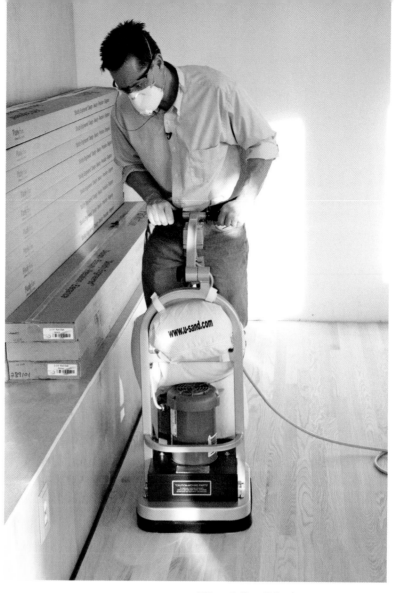

Although it's a little slower, you can use a multi-disk sander in place of the big machine, edger, and buffer.

There will usually be some prep work required before you can begin re-sanding an old floor. Pet stains, water stains, scratches, and dents may all require too much sanding to remove completely and require board replacements (see chapter 12). Don't sand cupped floors until the cause of the cupping (typically dampness from below) has been fixed and moisture testing indicates that the flooring is at an acceptable moisture level for the climate (see chapter 1). The moisture content of the top and bottom of the flooring should also be within 1% of each other. Correcting the moisture condition after the floors are sanded will result in the flooring flattening, which will create a crowned appearance because the board edges were sanded down.

Check for loose boards and refasten if need be. Loose boards can make the sander create waves and chatter across the floor. Too much flooring movement can also make the finish flake or peel along board

Take Lead Seriously

Most people are at least vaguely aware of the dangers of lead-based paint (banned by the federal government in 1978), and it's no big leap to realize that painted floors may contain lead. Fewer people realize that many varnishes once contained lead. So, anytime you're working on floor not sanded to bare wood since about 1980 (this assumes a two-year shelf life for paints and varnishes), you should test for lead. Most paint stores have lead test kits. If you find lead, it may be a good idea to turn the job over to a pro.

For many years, the EPA required contractors to notify homeowners of the dangers of lead dust, but the requirements are about to get a whole lot stricter. As of this writing, the EPA was implementing certification requirements for contractors working in houses built prior to 1978 that *may* be frequented by pregnant women or children younger than six. As a practical matter, this affects every house built before 1978. The new rules expect to take effect on April 22, 2010. In a nutshell, any time an area of greater than 6 sq. ft. *potentially* containing lead-based paint is disturbed, the contracting firm must be registered and at least one EPA-certified employee must be on-site.

Warning signs must be posted, the occupants given a brochure on the hazards of lead, and dust-containment measures such as enclosing the contaminated area with taped-off plastic sheeting and sealing off ducts in the work area must be undertaken. Anyone working in the area should wear a NIOSH-approved respirator with N100-designated cartridges. Additionally, any sanding equipment used must be equipped with dust collection and a HEPA filter. HEPA filters catch 99.97% of airborne particles of 3 microns. This may be a challenge for flooring contractors, as most sanding equipment uses 5-micron filtration.

There will also be cleaning, verification, and record-keeping requirements. This is only a partial listing of the requirements, which will likely change over time. For current info, visit www.epa.gov.

edges. With access from below, you can use screws to secure loose boards. Just be sure they aren't long enough to stick through the sanded flooring. Alternatively, add top nails.

WAXED FLOORS

Most clear floor finishes can be removed by sanding, though if the flooring has been waxed, you may have to rethink your approach. Normally, I can tell a waxed floor as soon as I see it because of the presence of white water stains. You can determine if a floor's been waxed by scraping it with a fingernail. If a gray, gooey substance comes up, it's wax. You can also place a drop of water on the floor. If the finish becomes cloudy, it's wax. Sanding waxed floors can clog up the paper in seconds. The key is to minimize the temperature of the abrasive and keep the wax from

melting. First, try a test pass with a 36-grit belt on the big machine. Back off the pressure on the drum and sand at a fast walking pace. If the paper clogs, change it and try walking faster. Do not try to overlap your passes the normal 2 to 3 in. Use full fast passes to skim off the wax. You may have to sweep up the debris, as the vacuum on the sander will not pick it up properly.

For the initial sanding of previously finished floors, move quickly to avoid overheating the abrasive and melting the finish. Sanding at a slight angle aids in aggressive stock removal when flattening is necessary.

When sanding the perimeter of the room, test with 36-grit paper on an edger and move quickly. The moment you slow down, heat will build up and the wax will clog the paper. I prefer to use my Festool 150 EQ in its aggressive mode with Festool's 40-grit Crystal paper. Festool's superior dust extraction keeps the abrasive cooler and cleaner.

Even after sanding, the new surface finish may not adhere properly. Wax residue is impossible to remove from the gaps between the boards and the finish may peel on the board edges. Some solvents will dissolve

Protecting an Occupied House

Finishing floors in a house under construction is less work than in an occupied one. Even though all floor-sanding equipment has dust collection, it's inevitable that some dust will escape. Closing off the rest of the house from the room where I'm working not only helps contain the dust, but it's also always a great diplomatic move that lets the homeowner know I care about protecting their home.

Even with dust collection measures on the sanding machines, closing off the rest of the house with thin plastic is a good idea to localize the dust and keep the homeowner happy.

Edging a resinous pine floor calls for coarse abrasives and fast movement to prevent clogging.

wax, but using them on flooring is counterproductive. The chemicals will drive the wax farther into the cracks and after you coat the floor with finish these chemicals and the wax will rise to the surface and lift or peel the finish. Even if the floor is completely sanded, the wax remains in the cracks between the boards. Trowel filling the floor may help lock in the contamination. Trowel filling is similar to filling nail holes, but on a floor-wide scale. Use either a commercial wood filler, or better, filler made from sanding dust and wood flour cement. The filler should be thinner than the putty you'd use for holes or spot filling, and it's applied by working it back and forth across the grain until all cracks are filled. Apply the filler using a trowel or drywall taping knife. Allow filler to dry, and then give the floor its final sanding.

Once a floor has been waxed, the safest thing to do is use only wax finishes on it in the future. That said, I've probably refinished 1,000 floors that had been waxed without a single failure. Most other finishes sand off readily, though penetrating oil finishes may require more stock removal than film finishes.

PAINTED FLOORS

Painted floors may also require an additional step. Depending on the thickness and condition of the paint, it may be necessary to strip it before sanding. Thick layers of paint will clog sandpaper. You should also test paint for lead prior to stripping, and, in fact, contractors are now legally required to do so by federal law. If the paint contains lead, you may want to consider hiring a lead abatement contractor instead of doing it yourself.

Sanding Prefinished Flooring

Sales of prefinished flooring now exceed sales of unfinished flooring, but like any floor, prefinished floors eventually require refinishing. Many prefinished floors use an extremely tough aluminum oxide finish, the same mineral used in sandpaper. Sanding these aluminum-oxide finishes is a hot issue in the wood flooring industry. Be aware that the sanding dust from aluminum oxide finishes is very dangerous to breathe, so it is essential to wear a respirator when sanding.

Sandpaper with ceramic abrasives, not aluminum oxide, performs the best on prefinished floors because the abrasive grains fracture into smaller razor-like pieces more times before the abrasive dulls. Finer paper, such as 80 grit, works better than coarse paper.

Prefinished floors have beveled edges, and re-sanding may remove the bevels. And if you don't sand out the bevels, and don't sand consistently, they may end up being uneven. If you don't remove the bevels, use a scraper to remove the old finish from them. Be aware that finish may pool in any bevels as you coat the floor.

When stripping prefinished flooring, getting the old coating out of the bevels requires the attention of a scraper.

CHAPTER ELEVEN

Finishing Wood Floors

BETWEEN LAYING THE FLOOR AND SANDING IT, or just re-sanding an old floor, the amount of effort it takes to have reached the point where you're ready to apply finish is substantial. You're done with the heavy work—no more slamming a mallet into a flooring nailer for hours on end or driving a throbbing 200-lb. sander across the boards. But having done all this work, you want to be sure the finish you use will provide the effect you want, and you want to be sure that you know how to apply it correctly so that it lasts for decades.

Finishes are protective coatings that sit on or in the surface of the wood. They enhance the

Modern finishes and techniques provide options for both pros and DIYers to lay down great-looking floor finishes.

appearance of the wood and impede moisture absorption into the flooring, which helps limit dimensional changes. Finishes seal the pores of the wood, keeping out dirt and making it easier to clean. Finishes also help protect the underlying flooring from wear and scratches.

Not all newly sanded floors get coated with finish immediately. Often, staining or dyeing the wood occurs first. Stains and dyes color the wood, but they have some fundamental differences discussed later in this chapter. You don't need to color the entire floor, either. I frequently use stains and dyes to accent particular areas. You can also use stencils with stains, dyes, or paints, as well as appliqués that go on like decals. All these decorative effects end up coated with finish to protect them from wear, but some, like paints, go on between coats of finish.

Flooring Finishes

I work with a variety of finishes that include wax, oil-varnish blends, conversion vanish, and urethanes. The urethanes are further divided into oil based, waterborne, moisture cured, and UV cured. Some are easily applied by homeowners, but a few require special equipment and skills that are best left to the pros.

Flooring finishes are unique blends formulated to make a finish with the qualities the manufacturer wants. Resins polymerize into the hard wear surface we walk on, and solvents thin these thick, gooey substances to a workable consistency. Alkyds improve adhesion, and silicates act as a flattening agent to make the finish less shiny. Anti-settling agents keep the flattening agent from separating out. The resins in floor finishes have high surface tension, which can prevent proper flowing and leveling of the finish. Leveling and wetting agents lower the surface tension of the finish so it flows uniformly over the floor surface before it dries. Waterborne finishes contain small amounts of surfactants and emulsifiers that can produce bubbles. Added antifoaming agents break these bubbles and stop them from producing imperfections. The list of additives goes on and on, and the addition of each requires the adjustment of another. Because of the complexity of these formulations, I'm careful to avoid additional additives not specifically approved by the finish manufacturer.

Wax Finishes

I love the look of a hand-rubbed wax finish. It is not quite French polishing, but it does provide beautiful warmth and depth. I have only one waxed floor in my home. It is in the smallest room and made from a black walnut tree that I milled into my own plank flooring.

Wax finishes are generally applied over a stained or sealed floor. I apply the wax sparingly by hand and polish it with a buffer. Too much wax may cause scuffs and attracts dirt. As the floor is buffed, the wax softens and penetrates the wood. Wax finishes are very forgiving to apply but wear quickly. And, of course, wax can be so slippery it will keep a child occupied for an hour sliding across the floor in socks.

1. Form a ball of paste wax about the size of a golf ball. You'll need to apply two coats and buff between them.

2. Wrap the paste wax with cheesecloth or terrycloth. Make sure the cloth is porous enough to allow the wax to flow out evenly.

3. On plank floors and strip wood flooring, rub the wax on in the direction of the flooring. On pattern floors, rub the wax on in a circular motion to prevent streaks. Liquid wax would be rubbed in the same way as paste wax.

4. Use a white pad or a steel wool pad to buff in the wax. The sheen level of the floor depends on the polishing pad or grade of steel wool that is used. The 3M white pad used here will give the floor a semi-gloss sheen.

5. Allow the wax to dry for one to two hours before polishing it to the desired sheen. Make sure you polish each coat to at least the desired sheen level. You can always make the floor less glossy on the second coat but not the other way around.

Wood floor wax comes in paste wax or liquid (the author prefers paste). Sand the floor normally and finish with a stain or sealer. If using a sealer, buff after drying with a 3M abrasive maroon pad.

Each finish has its virtues. Wax or an oil-varnish blend such as Waterlox® (www.waterlox.com) looks great on a 200-year-old wide-plank floor. Oil-based urethane (I prefer Fabulon®) is perfect for a strip floor neglected for 20 years that needs a slight ambering effect to hide minor imperfections. Two-part water-based urethanes have few volatile organic compounds, or VOCs (see the sidebar below), fast dry times, and are durable, making them good choices for an occupied house. The latest advance is a finish that cures instantly with ultraviolet light. It has a lower VOC level than the other finishes but is extremely durable and chemically resistant—great for occupied homes or businesses that need quick access after floor refinishing.

In addition to the choices in finishes, methods of application vary. There is some overlap, but, generally, oil-based and water-based finishes are applied differently.

Personal Safety

Volatile organic compounds, or VOCs, are the evaporative solvents that give finishes their chemical smell. Exposure to many of them is associated with short-term health issues such as nausea, headaches, and contact dermatitis. Worse, long-term exposure can lead to asthma, cancer, and other chronic health problems. VOCs liberated from finishes can enter the body through the lungs, eyes, and skin. Proper respiratory, eye, and skin protection should always be used when working with floor finish.

Every finish manufacturer makes available a Material Safety Data Sheet (MSDS) for each product. These can often be found at the manufacturer's website, and they'll tell you the level of personal protection required. At a minimum, wear a respirator with cartridges rated for

VOCs. Some finishes are worse than others, and when I use those containing xylene or xylol, I wear an industrial respirator that covers my full face and my eyes, as well as gloves and a Tyvek® suit.

In addition to the immediate health hazards of VOCs, they also play a role in overall air quality. VOCs react with other compounds in the air to form ozone. While ozone in the upper atmosphere does us good by absorbing harmful UV radiation, at ground level it's a pulmonary irritant that can induce asthma and aggravate chronic lung conditions. Because of this, the allowable level of VOC off-gassing from finishes has been restricted in a number of states. Depending on where you live, this may affect the formulation and availability of some finishes.

Personal protective gear is a must when finishing. At a minimum, a close-fitting respirator with cartridges approved for VOCs is necessary.

A CLEAN FLOOR IS A SMOOTH FLOOR

Nothing causes imperfections in a new finish quicker than improper cleaning. Finishes magnify any specs of dust or dirt and make them look like mountains. Dust can also make the finish bubble or impede adhesion.

It's not just the floor that needs cleaning before finishing. Dust from the ceiling, the walls, the windowsills, and the lights will fall on your wet finish. Start by vacuuming every inch of the room, from the ceiling down. Allow half an hour or so for any missed dust you disturbed from these surfaces to settle, and then vacuum the floor. Vacuum inside the HVAC vents and carefully vacuum the fins on any radiators. Move the vacuum head from dirty to clean. Vacuuming at an angle to the flooring will help remove dust between boards. Vacuum the perimeter of the room last.

You're not done yet: Next, you have to wipe the floor with "tack cloths." In the past, flooring professionals would tack the floor with clean, lint-free rags lightly dampened with the solvent used in the finish. For oil-based finishes, they would use mineral spirits, and for water-based ones they would use water on the tack rag.

Times have changed, and I prefer to dry-tack the floor with micro-fiber tack cloths, which are readily available from finish suppliers, paint stores, and home centers. Each one of the micro fibers is so small that a 1-in. by 1-in. piece contains about 3 million feet of fiber, which

After vacuuming the walls, ceiling, and trim, thoroughly vacuum the floor. Start vacuuming at an angle to the boards as this does a better job pulling dust from between them. Finish by vacuuming with the direction of the boards.

ABOVE LEFT Tack the periphery of the room first, and then work toward the center. Always pull the tack cloth in one direction so that debris doesn't fall off when the direction is changed.

ABOVE RIGHT Tack the floor with five separate, clean cloths to remove all the dust.

Clothes and shoes can contain debris that could fall on the floor during finish application. Some contractors wear dust-free booties over their shoes and Tyvek painter coveralls when finishing.

makes lots of space for dust to accumulate. Micro-fibers have a positive electrical charge. Dust and dirt have a negative charge. The positively charged micro fibers draw the negatively charged dust particles like a magnet. Dirty micro-fiber cloths vacuum clean or wash out.

I like to tack with five separate cloths, meaning that I go over the floor five times, each time with a pristine tack cloth. This may seem like overkill, but the dust keeps coming up. As a last note, never use an automobile tack cloth because it may contain silicone. If you get silicone on the floor, the finish won't adhere but instead will tend to crawl away from the contaminated spot. The only fix is re-sanding.

WORK IN THE PROPER CONDITIONS

Apply most finishes when the temperature of the work area, floor, and finish are between 65 and 80°F and the relative humidity is between 40 and 60%. When it's too cold or too humid, the finish may take a long time to dry, giving dust more time to settle on the tacky surface and slowing down the time before additional coats can be applied.

If it's too hot or too dry, the finish can dry before the marks from the applicator have a chance to level out. It's important for the finish itself to be no hotter than room temperature. If it's been sitting in the sun in your truck, the finish can be as hot as 165°F. If you apply finish at this temperature, it can skin over quickly and trap bubbles. And watch out for sunny spots, particularly on dark stained floors. These areas can get hot enough that finish dries before it has a chance to penetrate the wood, which interferes with adhesion and may cause blistering. Windows and doors that flood a floor with sunlight should be shaded long enough for the floor to be within the finish manufacturer's recommended temperature range.

It may be possible to work in hot and dry conditions by adding a retarder to the finish. Thinning the finish with a solvent like mineral spirits or water, depending on if the finish is oil or water based, might help if the directions say that's okay. However, finish formulations represent a balance between varieties of substances. If a manufacturer makes a dedicated retarder for their product, that's what you should use.

If the HVAC system is forced air, turn it off in that room until the finish has time to completely flow out, meaning that the applicator marks have settled to a level surface. Air blowing across the floor can speed the evaporation of solvents, allowing the finish to harden before the applicator marks have leveled. Usually this takes about half an hour.

Because waterborne finishes literally flood the floor with water, take moisture content readings of the flooring prior to coating. The floor is generally ready for the next coat when its moisture content is within 1% of the original reading.

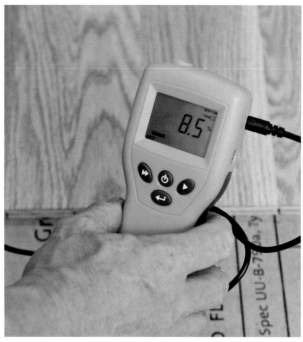

Check the moisture content of the floor between coats of waterborne finish. Re-coat when it's within 1% of the reading before finishing.

WHY USE A SEALER?

Many people, including some flooring professionals, dismiss the need for first-coating a floor with a sealer, but sealers have so many advantages that I always use them. Sealers are generally less expensive than the main finish. The sealer coat doesn't require the same qualities as the final coat of finish: It won't bear traffic, so you can use a less durable finish based on a cheaper acrylic resin. Most sealers are fast drying, which generally allows you to apply the sealer and the first coat of finish in the same day.

Water-based finishes can raise the grain and don't penetrate wood very well. For this reason, most manufacturers make sealers to be used

First-coating a newly sanded floor with sealer promotes better adhesion, reduces the chances of tannin staining, and, because sealers dry quickly, allows two coats in one day.

Side bonding occurs when the finish glues several boards together and they shrink as one.

Even if you want a satin or semi-gloss finish, use gloss finish for the initial finish coats. Gloss finishes are clearer because they lack the flattening agents added to satin and semi-gloss finish. The finish will be clearer if only the surface coat contains flattening agents.

in conjunction with their water-based finish. These sealers are generally solvent based, which limits grain raising so less sanding is required to smooth the floor again. Water-based finishes can dissolve the tannins in oak and other tannin-containing wood species. This shows up in the final coat as a green tinge, which can be particularly objectionable in a white pickled floor. Because sealers keep the water from reaching the tannins, they minimize the risk of discoloration.

When top-coating with water-based finish, it is important to use a sealer approved by that finish manufacturer. For example, oil-modified urethane can take as long as 30 days for all the solvents to evaporate. Coating such a finish with water-based finish, which may cure fully in a day or two, traps the remaining solvents from the oil-based finish and can lead to adhesion failure.

Side bonding occurs when finish flows between the floor boards and glues them together. This becomes a problem in times of low humidity when the boards shrink. Rather than the shrinking manifesting itself as small, unobjectionable cracks that appear between multiple boards, the side-bonded boards shrink as one wide board. When they reach a board that isn't side-bonded, all the shrinkage from the side-bonded boards appears as one large crack. A good sealer gets between the boards first, making a weak bond that will fracture as the boards shrink. Sealers such

The Importance of Ventilation

Finishes don't cure simply because their solvents evaporate. They cure when their resins, which are relatively short organic molecules, link together, or coalesce, because of a reaction with an added catalyst or because of oxidation. However, coalescence can't happen until the solvents evaporate, so evaporation is a key to the finish curing.

If the room where the finish has been applied isn't well ventilated, the initial evaporated solvents tend to saturate the air near the floor and inhibit further drying. If you don't ventilate the room, the finish will cure slowly and may never fully harden. But neither do you want the finish to dry too quickly initially, or it won't have a chance to level out.

The key is to limit air movement for the first half hour to give the finish leveling time. Then for the next four hours or so, you should ventilate the room as much as possible while keeping it within the recommended temperature range. To limit dust contamination, it's best if you can draw the air from the room rather than blowing into it.

as Synteko's® Sealmaster advertise a "high crystalline structure" for just this reason. The sealer is slightly brittle and allows this bond to fracture.

The resins in oily woods such as teak, rosewood, and Brazilian walnut can prevent oil-based finishes from adhering or curing. Always seal these woods within two hours of sanding, or the resins can migrate up to the fresh surface. A sealer acts as a barrier coat protecting the oil-based polyurethane from the wood's resin. Sealers dry quickly, minimizing the chance that the resins can migrate through them.

SHELLAC One of my favorite sealers is one of the oldest. Shellac has been used as a wood finish for about 3,000 years. It comes from a resin secreted by the Asian lac beetle. The resin itself is food-safe and is used to create a shiny finish on some candies. Shellac adheres tenaciously to wood and provides great adhesion for other finishes. The main caveat is that it naturally contains wax that can cause adhesion problems with urethanes. Using dewaxed shellac avoids this problem. Dewaxed shellac tends to gum up when sanded, but you can coat the shellac without sanding and sand the next coat.

Shellac dissolves in ethyl alcohol and dries very quickly. It is non-yellowing and will not darken with age like oil-based finishes. Unlike other finishes, you can apply shellac in temperatures as low as 40°F.

One final word on sealers: Apply them using the same methods as other finishes.

Choosing a Finish

Oil-based finishes, whether made from old-style varnish resins or modern polyurethane, were the standard flooring finish until the 1990s. About then, ever-tightening VOC regulations coupled with improved formulations started to give water-based finishes a leg up. Indeed, in some states (California being the leader), many oil-based finishes can no longer be found.

Building traditions die slowly, and truckloads of oil-based finishes are still available to purchase. One reason is their look. Water-based finishes tend to dry clear, while most oil-based finishes lend an amber tone to the flooring. Additionally, oil-based finishes usually take longer to dry and cure than water-based finishes. Although that can be an advantage, in arid conditions water-based finishes can dry before the applicator marks in the surface have a chance to level out. This is less of a problem with slower drying oil-based finishes.

OIL-VARNISH BLENDS

Oil-varnish blends are some of my favorite finishes. They penetrate deep and seal wood fibers beneath the surface, and are one of the few finishes that act as their own sealer. Unlike urethanes, which can look like a sheet of plastic from the way the finish lies on top of the wood's surface, oil-varnish blends create the look of rich, hand-rubbed patina that enhances the grain of the wood.

Most oil-varnish blends incorporate either tung oil or linseed oil. Tung oil comes from tree nut, and linseed oil comes from flax seeds. The oil is mixed with one of several varnish resins, and often wax is added. Tung oil is considered superior to linseed oil because the molecules are smaller, thereby allowing it to penetrate the wood. Linseed oil yellows when exposed to UV light, while tung oil does not. My favorite oil-varnish blend is Waterlox, which is tung-oil based and contains no wax. Because it contains no wax, there's no problem recoating Waterlox with other finishes in the future.

Oil-varnish blends adhere well to hard, oily woods like teak, Brazilian cherry, and Brazilian walnut, where many finishes fail. They remain flexible, suiting them particularly well for softwoods such as pine and

Oil varnish creates a low luster sheen that's easily renewed.

fir. Harder finishes such as waterborne urethane used on soft woods can develop white marks if the softwood is dented. Scratches in urethane coatings have distinct edges that collect dirt and are very visible. Repairing such scratches often requires re-sanding the entire floor because you can't feather in a small repair with urethane. Because oil-varnish blends produce a minimal surface build, scratches often go unnoticed and can be repaired simply by applying more finish. No sanding is required between coats of oil varnish.

CONVERSION VARNISH

Conversion varnishes are also known as Swedish finishes. Applying them requires more skill than other finishes because they dry fast and are unforgiving of application mistakes. I don't recommend them for DIY use. For pros though, they can be the right choice in some circumstances. I've used Synteko Classic acid-curing conversion varnish in hair salons due to its high resistance to chemical spills. It is extremely durable and a good candidate for homes with pets. It is critical that proper respirators, eye protection, and skin protection be worn during application; the finish vapors are flammable.

OIL-MODIFIED POLYURETHANES

Oil-modified, or oil-based, polyurethanes are made by reacting vegetable oils like linseed, soybean, or safflower with polyhydric alcohol and di-isocyanate. This reaction forms a compound with long polymer chains known as polyurethane, or simply urethane. Oil-modified urethane (OMU) is never perfectly clear and tends to amber in color with age. This is especially true with linseed-based urethane. Soybean oil is slightly lighter and darkens less with age. Safflower oil darkens the least and is a good choice for lighter wood such as birch and maple. Polyurethane made with safflower oil is sometimes sold as non-ambering.

The finish wears well and is relatively easy to learn to apply, with a longer working time than many other finishes. Small mistakes are easy to repair during application. OMU takes three to seven days under optimum conditions to cure hard enough before furniture can be placed back onto the floor and about two weeks for area rugs.

The downside to OMU finishes is the VOCs. In a confined area, it can be difficult to purge all the solvents from the air unless the windows are opened. As the solvents build up in the air, the curing process of the finish slows. I have had many jobs where one coat of finish took three days to cure properly. Wear a respirator when working with this finish to protect your lungs (though the solvents can still be absorbed through your eyes). The smell of OMU finishes can linger for days or even weeks. Homeowners who suffer from allergies or are sensitive to smells may find this objectionable.

Oil-modified urethane finish is one of the most widely used floor finishes. Its high VOC levels, however, are making it harder to obtain. It has a longer working time than most finishes and gives floors furniture-like quality. It brings out the rich colors of wood such as walnut and ambers over time, which helps mask any imperfections in a floor.

WATERBORNE URETHANES

Waterborne urethane has the urethane resin suspended in water. Although most contain a small amount of solvent, they have low VOC levels. Water-based urethane dries in less than a third of the time of most solvent-based finishes. Under optimum conditions, furniture can be replaced in about three days. Wait about one week for area rugs.

Water-based finishes are clear and do not bring out color in woods like walnut. They are considered non-ambering. Water-based polyurethanes are harder than oil-based urethanes, and some consider the finish to have a plastic look. Scratches can have a pronounced white color. These finishes require additional learning and specific application techniques, which differ from those used with oil-based polyurethane. One drawback is that water-borne polyurethane can and often will raise the grain of wood if you do not seal the wood first.

TWO-COMPONENT WATERBORNE URETHANES

Two component urethanes like Bona® Traffic (www.bona.com) and Synteko Best (www.synteko.com) are non-ambering, extremely durable water-based polyurethanes. As with other waterborne urethanes, these also have some VOC solvents. Both brands use a poly-isocyanate to initiate the molecular cross-linking that speeds the resins' hardening. Once the cross-linker is added, you have about four hours to use the product. Both finishes outperform most one-component waterborne polyurethanes.

While the cross-linker speeds the curing process, it happens in concert with the solvents evaporating. The water is supposed to evaporate first. If the organic solvent evaporates first, the finish won't form a film. Most of the time, this isn't a problem but it can be when the atmospheric relative humidity is high enough to slow the water's evaporation. When it's very humid, ventilate sparingly to give the organic solvents time to evaporate first.

Oil-modified urethanes have long working time, making them relatively easy to apply. They do off-gas many VOCs, and the smell can take weeks to dissipate fully.

Applying Oil-Based Finishes

Because of their long drying time and good leveling, oil-based finishes make it easy for the novice to achieve near perfect results. They are traditionally applied using a 4-in. China bristle brush to first cut in the edges and a lamb's wool applicator to coat the field. Dip the applicator in the finish and run from wall to wall with the wood grain. It's important to avoid stopping in the middle of the room because wherever the applicator stops, the flattening agent concentrates, forming a cloudy mark in the finish. When the room is too big to make it from wall to wall on a single dip from the finish bucket, pour a line of urethane poly the length of the room and run the applicator along that line. At the end of the room, lift the applicator from the floor in one smooth motion.

That said, I no longer use lamb's wool applicators because they tend to shed small fibers in the finish. Instead, I use rollers for the center of the floor and cut in using a china bristle brush.

No matter how you apply the finish, the initial steps are the same. Stir the oil-based finish in its container—don't shake it, as shaking makes bubbles. The flattening agents in urethane tend to settle to the bottom, so I stir with an up-and-down as well as a circular motion. Pour the finish through a strainer into a dishpan lined with a clean garbage bag. The garbage bag makes a dust-free container every time and aids in cleanup. Allow the finish to sit until the bubbles escape, and then apply. If the area will require more than one can of finish, mix them together first to ensure a uniform gloss.

The traditional tools for applying oil-based finishes are a lamb's wool pad and a China bristle brush.

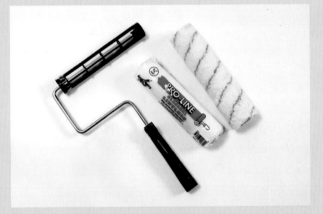

The author prefers rolling on all finishes, using solvent-resistant roller covers for oil-based finishes. Look for roller frames with straight ribs that won't distort the covers.

WATER-BASED ACRYLIC

Water-based acrylic finish is the least expensive floor finish. It looks like all water-based products but is considered the least durable. Given the amount of prep work it takes before finishing a floor, I don't think the savings are worth it. That said, there are some acrylic-based sealers that are fine to use. Sealers don't have to be very wear resistant because they get coated with more durable finishes.

URETHANE EMULSION

Most oil-modified urethane resins use mineral spirits as a solvent. Emulsions use no mineral spirits, and the urethane suspends in water. When the water evaporates, the urethane joins together to form a surface film. Cross-linking occurs through a reaction with oxygen from the air.

Urethane emulsions begin to cure quickly as the water evaporates. As with many other finishes, it's important to limit airflow until the surface flows out level, and then to provide good ventilation to encourage curing. These finishes provide a hard finish with low VOCs.

MOISTURE-CURED URETHANE

Moisture-cured urethane reacts with minute traces of moisture in the air, cross-linking the finish into a tough, resilient coating. The finish provides excellent abrasion resistance and good chemical resistance. Moisture-cured urethanes may harden slowly in extremely dry conditions, allowing them to soak into the wood and not build a film. Extremely high humidity can create problems as well. Moisture-cured urethane off-gasses carbon dioxide as it cures. Very humid conditions cure the surface too quickly, which in turn can trap carbon dioxide bubbles. It's best to apply moisture-cured urethane when the relative humidity is between 30 and 70%.

I personally avoid moisture cure unless the customer fully understands that I need to dress up in an air-fed space suit during application. I want to protect lungs, eyes, and skin from the xylene and other solvents that moisture-cured urethane off-gasses. I also tell my customers not to use their home until I properly ventilate it. The vapors are flammable.

ULTRAVIOLET-CURED URETHANE

Ultraviolet (UV) light has been used industrially to cure coatings in factory settings for over three decades. Recently, site-applied UV coatings have made their way into the floor-covering markets (www.uvcuredfinish.com). First, it was concrete slabs, then vinyl tile, and now wood flooring.

UV finishes contain photo initiators (PI) that react with UV light to form free radicals that cause the finish to cross-link. UV finish is very

UV-cured urethanes harden instantly with exposure to UV light. This professional finish allows the room to go back into service immediately.

durable, but its greatest attribute is its instantaneous cure. UV finishes allow homes and businesses to be fully usable right away. UV equipment used for site-applied finishes consists of a lamp assembly attached to the front of a mobile cart. As the machine moves, a shutter opens up that exposes the floor to UV light, curing the finish.

Anyone working in close proximity to UV light must wear UVA/UVB-blocking polycarbonate eyeglasses, long-sleeve shirts, long pants, and SPF-45 sunscreen on any exposed areas of skin. On oily wood species, the UV-curing process can draw pitch, sap, and oils to the surface of the flooring possibly diminishing adhesion. Prior application of a sealer will prevent this.

Rolling on a Finish

Rollers are my preferred way to apply both oil-based and water-based urethanes. I use roller covers that do not readily break down from solvents for oil-based finishes, and covers made from synthetic materials for water-based finishes. I generally use a ¼-in. nap roller for oil base and ⅜-in. nap roller for waterborne finishes. (Another alternative when coating large areas with waterborne finishes is to use a T-bar applicator, as shown on p. 254.)

One of the keys to any finish application is to keep a wet edge. This means that you should always be able to tie coated areas together by rolling, brushing, or padding the finish onto an area that has not yet dried or become unworkably tacky. Particularly with fast-drying water-based finishes, you have to work quickly and not coat an area whose edge will dry before you can blend it with the edge of the next area you're coating. There's no formula for this, as it depends on the finish and the conditions of the day. Until you know what you're doing, it's a good idea to work relatively small areas.

The first step is to coat the edge of the floor, cutting into the wall. For speed, you can pour a line of finish along the floor first, followed with a pad (as shown in the photos on p. 253). One great advantage of rollers is

1. Use a clean watering can to pour out a metered amount of waterborne finish along the edge of the room. The small-diameter spout makes it easy to control the flow.

2. Spread the finish evenly with a synthetic trim pad next to the walls and tight areas. When you begin rolling the floor, go over this area as well to ensure a uniform film thickness.

3. Rollers don't have to work with the grain, a great advantage at the end of a floor.

4. Pour the urethane directly on the floor, in a paint tray, or in a line down the room. All methods work.

5. Roll the puddled urethane out, working in one direction.

6. After laying down the finish initially, move the roller a couple inches into the wet finish and pull it all the way from wall to wall to tie in any ridges and stop marks.

Using a T-bar Applicator

One way to coat large areas quickly with a water-based finish is to use a T-bar. T-bar applicators use a snow plowing method to spread the finish. The edges are cut in with a pad, as when rolling. The ends of the floor need to be pad-coated out for a foot or so to provide room to turn the T-bar.

T-bar applicators (see the top left photo below) use a synthetic pad on a bar. The bars come in different weights to meter the finish thickness—heavier ones put on thinner coats. The finish manufacturer will specify which weight to use.

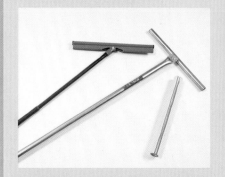

1. Pour a line of finish down the length of the room, and then pull the T-bar in a snow plowing or squeegee motion, displacing the extra finish onto the area to be coated next. When there's not enough finish for the next pass, pour more on the floor.

2. Turn the corner without stopping and keeping steady, even pressure on the applicator.

3. At the end of the room, keep the applicator moving in one direction but make a 180° turn, removing the excess finish away from the wall.

4. After the turn, press down on the back of the T-bar to remove any extra finish.

5. Blend the area where the curve was made, and begin the next pass.

that you can apply the finish with or across the wood grain, though it is still best to run from wall to wall when applying finish. This is very helpful at the ends of a floor, in cut-up rooms, or with parquet flooring. Rollers may cause small bubbles; take care not to overwork the finish—just roll it until the finish is even and move on.

To avoid stop marks, it's best to work the entire length of the floor with each pass. Plan your exit. When you approach the last wall, stop rolling and cut in the edge. Then close the gap between the cut-in edge and the body of the floor so that you'll have room to walk the final pass without stepping in the finish.

Recoating a Finish

There are two kinds of recoating: Building up additional coats on a new finish and renewing an existing finish. Recoating new floors is straightforward. Most floor finish manufacturers recommend two to three coats of finish, which provides somewhere between 0.003 and 0.004 in. of sur-

ABOVE LEFT Use a 3M Scotch-Brite pad with abrasive strips to sand between coats. Use 180-grit abrasive strip for oil-based polyurethane and many water-based sealers, and 240-grit abrasive strips for waterborne finishes.

ABOVE RIGHT Abrade by hand where the buffer can't reach using a maroon Scotch-Brite pad.

face film. It is imperative that each coat properly adheres to the previous finish coat.

Recoating a new finish may require a small amount of abrading, or even none. Oil-based urethane generally requires abrading to create small scratches (tooth) for the next coat to adhere to. Most waterborne finishes can be recoated within 48 hours without intercoat abrading, unless you have some raised grain or imperfections that need to be removed.

I buffer with a 3M abrasive Scotch-Brite® pad with abrasive strips to intercoat abrade my floors. In the past, flooring professionals used worn abrasive screens for this, but these screens can create scratches that are too deep and spaced too far apart. The new system provides a smaller, more plentiful and uniform tooth for the finish to adhere to. Be sure to abrade the entire floor evenly and abrade areas the buffer can't reach by hand. Vacuum and tack after abrading, and recoat according to the finish manufacturer's instructions.

RECOATING EXISTING FLOORS

Recoating old floors can be a little scary. No matter how much you test a floor there is no guarantee the new finish will adhere. The first step is to determine what finish is on the floor. Next, find out if it is contaminated with something that will cause problems with the coat of finish. Last, prepare the surface to accept a new finish.

Not all floors need recoating. If the floor is excessively worn, or has stains, dents, or permanent cupping, simple recoating may not provide satisfactory results. A simple test to determine wear is to take a well wrung out, damp towel and rub it over a worn area on the floor. While the floor is damp and glossy, it simulates the look of a new coat of urethane. If the water looks good, it's likely a new finish will too.

Try to determine what the existing finish is. It is important to find out as much information about the floor as you can. Are the homeowners the original owners of the floor, has the floor ever been recoated, and if so what type of finish was used? How was the floor maintained? What type of cleaners were used to maintain the floor and how often were they used? Overspray from furniture polish, cleaners that contain silicone, or wax on the floor can all cause adhesion problems.

You'll have to make sure the finish you want to use is compatible with the existing finish. Once you find out what that is (see the sidebar at right), the new finish manufacturer can say whether their product should work. It's a good idea to test a small patch of the existing floor to see if the new finish will adhere. This doesn't guarantee adhesion everywhere because there may be contaminants, but it's worth trying.

As long as the existing finish isn't wax, and the floor is in good shape, odds are you'll be able to recoat it. If it's a wax finish, the best thing to do is to renew it with more wax and a buffing.

Otherwise, the first step in recoating an existing finish is to clean the floor thoroughly. Vacuum first, and then clean the surface with a cleaner that is compatible with the recoating process. As with a new floor, the next step is usually abrading the surface of the old finish to create a mechanical bond or using a chemical system to create a surface the finish can bond to. Depending on the old and new finishes, you may be ready to vacuum, tack, and coat now.

How to Determine an Existing Finish

- Urethane finishes are resistant to mineral spirits or acetone. A drop of either will not affect the finish to any great extent.

- Lacquer dissolves in acetone in about 30 seconds with rubbing.

- Varnishes become gel-like a minute or two after applying acetone. Varnish also reacts slowly to denatured alcohol.

- Shellac dissolves quickly in denatured alcohol and turns into gel a minute or two after applying acetone.

- Oil finishes will absorb a few drops of linseed oil when rubbed in. If it beads up, the wood has a surface finish.

- Wax will turn white with a drop of water.

RECOATING PREFINISHED FLOORS Many new prefinished floors have some form of mineral suspended in the finish. These minerals make the finish wear longer, but they are often the same aluminum-oxide minerals used in sanding abrasives. Tying to abrade these finishes can be a challenge. The buffer tends to leave scratch or swirl marks, as some of the particles tear away from the finish and grind into the floor. Most prefinished flooring manufacturers recommend chemical bonding systems. Chemical systems either etch the surface or prime it to create a surface for the finish to bond to. Two such systems are Basic® Coating's TyKote® System and Bona's Kemi System.

Staining Floors

Almost everyone has stained a piece of furniture or some window trim, and it seems to be a simple task. This is probably the main reason why I have inspected so many failed stained floors. A tabletop is comprised of only a few well-selected boards. The boards are sanded using the same technique over the entire piece. A floor, on the other hand, may be made of thousands of boards from many trees. The boards have been cut at varying angles to the annual rings. Sanding the floor involves several types of sanding equipment. Each one of the sanding machines opens the pores of the wood in varying degrees, which affects how the wood accepts stain.

Water Popping

A technique called water popping can help make a stained floor appear uniform and darker. Apply water to the floor sparingly to open the pores of the wood. I prefer to spray the floor lightly with a pesticide sprayer. Do not walk on the damp floor or you'll compress the wood and change its ability to accept stain.

Allow the floor to dry fully before staining. If the pores of the wood are full of water, there will be no room for the stain. Waiting over night is generally sufficient. I take a moisture reading prior to water popping, and check it the next day. When the two readings are equal, the floor will be ready to stain. Be aware that water-popped floors accept more stain and so will require more time to dry after staining.

Pigment stains have relatively large particles suspended in an oil base and commonly create earth tones.

Other factors that affect how wood accepts stain include the species of wood, whether the floor was water popped (see the sidebar on the facing page), the stain type, and the staining technique. Because of all these variables, one container of stain can create a limitless number of shade variations. Wood stain is a type of paint with a low viscosity. The colorant penetrates the surface rather than remaining in a film on the top.

PIGMENT STAINS

Oil-based pigment stains are commonly used on wood floors. They have finely ground mineral powders held in suspension by some kind of thinner and mixed with resin binders. The pigments are added to an oil-resin vehicle to make this stain. The pigments are in suspension, which means they need constant stirring to maintain a uniform color.

Pigment stains work better with open-grain woods such as oak, ash, mahogany, and walnut. The stains settle into the pores and scratches of the wood's surface. They are good for highlighting the grain of porous woods, but they also highlight sanding imperfections. I have inspected floors from 10 in. away before staining and been unable to see any imperfection even with the use of a spotlight, yet sanding marks seemed to magically appear once the stain was applied. A successful stain job depends on meticulous sanding techniques (see chapter 10). Even with a perfect sanding job, the stain will not even out the dark and light natural tones of the wood. In fact, it may highlight them. Nonporous wood

Work with a helper to speed hand staining, and make sure to wipe off any excess stain before it dries. Line a plastic dish tub with a clean garbage bag for a dust-free stain bucket that's disposed of easily.

Preconditioning

Certain woods are prone to blotching when stained. Pine (particularly white pine), fir, hemlock, cherry, and maple are examples. One way to minimize blotching is to condition the wood. Stain manufacturers sell conditioners, but I find they dry too slowly. I prefer to use de-waxed shellac thinned to a ½- to 1-lb. cut, meaning that there's that weight of dry shellac dissolved in 1 gal. of alcohol. Premixed shellac is sold in 2-lb., 3-lb., and 4-lb. cuts, so thin accordingly. Shellac provides great adhesion, seals the pores uniformly, and is compatible with all stains and dyes except for those that are alcohol based.

species such as maple, pine, cherry, beech, birch, and hickory may stain non-uniformly and look splotchy. Preconditioning may help (see the sidebar at left).

The most commonly used stains for flooring are oil based. These stains are the simplest to apply and provide consistent and repeatable performance. Water-based stains are available but are difficult to apply without lap marks.

Perhaps the most important factor when staining is to remove all stain residue. Residual stain will interfere with finish adhesion, perhaps leading to peeling. For this reason, I only ever apply one coat of stain. I've never had a second coat of stain even out or darken the previous one enough to justify the additional risk of leaving residue. Darker colors can be obtained by dying the floor before staining it, finish-sanding with a rougher grit, water popping, or just using a darker stain.

Mix all the cans of stain to be used on a floor together to ensure a uniform color. If you're applying stain by hand, first wipe on a generous amount, and then wipe it off with a clean rag used in a circular motion. Follow by wiping parallel to the grain to make any unintended streaking less visible. Check the can for drying time, and allow plenty of it. If you're in too much of a hurry, you may end up having to sand the floor again when the finish starts to peel.

STAINING WITH A BUFFER

Staining a large area on your hands and knees with nothing but rags isn't a lot of fun. I much prefer to stain large areas using a buffer. It's better to use a self-dissolving stain—floor-specific stains will say on the can. Self-dissolving stains are made for staining large open areas without lap marks, which helps blend areas together. You'll probably have to go to a wood flooring supply house for such stains.

1. You can apply stain with either a scrap of carpet cut to the size of the buffer or a white buffing pad.

2. Pour about a half cup of stain into the center of the white pad or the carpet scrap. Turn the pad over, put the buffer on it, and keep going until you need to add more stain. Start the buffer away from the wall so it doesn't fling wet stain where it isn't wanted.

3. Apply the stain buffing quickly. Once a pass is complete, put a clean pad on the buffer and go over the same area to remove any residue.

4. Allow the stain to dry overnight and buff again the next day to ensure no stain residue is left on the floor to diminish finish adhesion.

Buffing on the stain usually uses half as much stain as the can says, but the floor may be a shade slightly lighter. I use either white buffer pads or carpet pieces cut to fit the buffer to apply the stain. Carpet samples work great and are cheap or free. All carpet yarn is manufactured from a staple or continuous filament fiber. Staple fibers are short, 6-in.- to 7-in.-long fibers spun together to make the yarn. Continuous filament fiber carpet works best for staining because it sheds fewer fibers. Polyester fiber carpet seems to be the best choice for buffing on stain because it has the correct absorbency characteristics. Polypropylene (Olefin) and nylon don't work well because they have a very low absorbency rate. Wool only comes as a staple yarn, which means it tends to fuzz. I avoid carpet treated with a stain guard because it might leave a contaminant that would interfere with finish adhesion.

Applying the stain with the buffer couldn't be easier or faster. Pour stain on the pad, and make a pass with the buffer. After that pass, change to a clean pad, and buff away any residue. To avoid any issues with finish adhesion, allow the stain to dry overnight and buff it out again with a clean pad the next day. I do the corners and edges by hand.

Dyes use smaller particles suspended in water or alcohol to lay down earth tones and create wilder effects.

STAINING WITH A LAMB'S WOOL APPLICATOR

You can also apply stain with a lamb's wool applicator, but I find that the buffer technique works better. It is very easy to apply too much stain with the applicator. If a lot of stain flows into the cracks in the flooring, you can have problems later. Stain that puddles between the boards takes a long time to dry, and may dissolve in the finish and bleed upward. Trowel-filling the floor is one way to avoid this. Another is to work as a team, with one person applying stain with the lamb's wool applicator and the other immediately wiping it off.

DYE STAINS

Compared with pigment stains, dye stains are made of much smaller particles dissolved into water, alcohol, or oil. Dye stains penetrate deep into the wood fibers and adhere better than pigment stains. Dye stains dry very quickly, making them difficult to apply evenly over large surfaces. Because they dry more slowly, water-based dyes are more forgiving than alcohol-based dyes. However, it's a tradeoff: Alcohol doesn't raise wood grain, though applying it requires more skill. I like using a concentrated dye solution called TransTint by Homestead Finishing Products® (www.homesteadfinishingproducts.com). It can be mixed with water or alcohol.

Dyes work better than pigment stains for close-grained woods such as maple or beech because the higher-viscosity stains don't penetrate the tight wood. However, a downside of dyes is that many are not as colorfast as pigment-based stains, particularly when they're in the direct sun. Many pros will follow a dye with a coat of pigment stain to improve colorfastness and to help even out the colors. The combination of the two provides an unmatched depth of color.

One common problem with dye stains is overlap marks. Streaks of darker color emerge where passes overlap. One way to avoid this is to

Ebonizing

Every once in a while I get a call to ebonize, or blacken, an entire floor. More commonly, though, I just add accent strips that simulate the look of expensive ebony. I start by dyeing the wood black and follow with a black or ebony pigment stain. Just using a pigment stain often leaves the surface looking a dark gray color, not the deep brownish black of real ebony. A coat of clear, non-ambering finish looks best over ebonized flooring.

Pickling

I have to admit that I'm happy that pickling, or staining a floor white, is no longer as popular as it was during the 1970s and early 1980s, but I do still occasionally get calls for it. To pickle a floor, I normally bleach the floor with wood bleach, followed with a coat of Dura Seal® oil-based Country White stain (www.duraseal.com). The floor has to be finished with a non-ambering finish or it will have an undesirable yellow tinge.

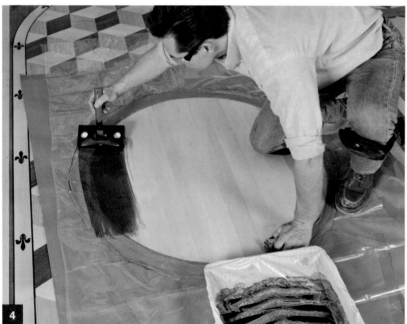

1. Dyes provide a great range and depth of color that make them great choices when you're going for a special effect. Begin by pouring water or alcohol into the mixing vessel. (The blue painter's tape burnished tightly to the floor and covered with plastic packing tape prevents the dye from bleeding through.)

2. Add drops of dye until you achieve the approximate desired depth of color.

3. Test the dye on a sample board that's been prepared exactly the same way as the floor before applying it to the floor.

4. Apply the dye with a pad applicator. Clear plastic packing tape that extends $\frac{1}{8}$ in. or so onto the wood keeps the dye from saturating the blue tape and maintains a crisp line.

5. For this small area, alcohol was used as a solvent to minimize grain raising. To minimize lap marks, the working area was almost flooded with the dye.

Accenting with Dye

Dramatic effects can be obtained with dyes. The oak and cherry in this unfinished floor blend, but an application of red dye creates a stunning effect.

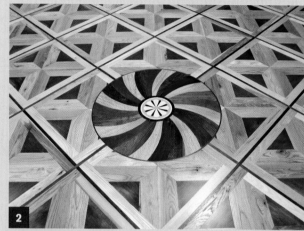

1. The parquet floor started out lackluster without color.

2. Adding just little bit of dye transformed the floor.

3. Burnish 3M painter's tape to the surrounding boards. Bleed wasn't a big worry here as the tape followed the edge of the boards.

4 Removing the tape left the cherry triangles vibrant and visible next to the oak crosses.

flood the flooring as quickly and thoroughly as possible. If you use a water-based dye, then water popping the floor first, letting it dry, and sanding off the raised grain before the stain goes on lessens the chances of ending up with a rough finish.

Faux painting allows limitless designs on wood floors. It can be freehanded or stenciled and is protected by the finish coat.

Gel stains work best for faux accents because they bleed less than thinner stains.

FAUX FINISHES

Paint, stains, and dyes can accent a wood floor or imitate inlays of more expensive materials. There are many great books on faux techniques, and I don't intend to do much more than provide the very basics here. For example, you should easily be able to find a hundred techniques for marbling and wood graining. I would do a great disservice if I tried to cover these topics in a couple pages. Instead, I will concentrate on techniques needed to add faux accents successfully to a wood floor.

Apply stains directly to bare wood. Gel stains work the best in preventing bleeding, which is when the stain or paint leaks past the tape or stencil. Acrylic hobby paints sold at craft stores work great for faux painting and are available in many colors. One surprising thing you can do with paint is to repair small sections of damaged floor. The wood floor should have two coats of finish applied before painting, and the finish should be lightly abraded with 220-grit sandpaper or a 3M abrasive maroon pad prior to faux painting. Coat all faux details with at least two coats of finish to protect from wear. The last important detail is to use a high-quality tape like 3M Scotch-Blue #2080 Painter's Tape and burnish the heck out of your sealing edge to prevent bleeding.

STENCILING One simple faux effect is staining a dark stripe on lighter wood. Carefully apply painter's tape in a pair of parallel lines and burnish its edges. Brush on a gel stain of the selected color, wipe off any excess, and then pull the tape to reveal a faux inlay.

You can make fancier shapes using a similar technique, as shown on p. 269. Pick an object you'd like to stencil your floor (I used an oak leaf). Burnish down an area of tape, and trace the outline of the object on it. Use a sharp X-Acto® knife to cut out the stencil. This is time-consuming,

Faux Repair

Damage to a finished floor can seem catastrophic when the only option that occurs to you is to replace that section of floor. But, often enough, a tearout or missing chunk can be filled in and painted to match the surrounding floor. Depending on the finish, you may need to recoat the entire room, but that beats tearing up the floor.

Carefully mask the damaged area, and blend acrylic paint to color-match the damaged strip in the floor.

A little paint saved the floor from having a new accent strip surgically installed.

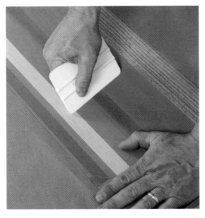

1. Burnish 3M blue painter's tape tight to the wood to prevent paints and stains from bleeding. You can use auto body spatulas, the back of a spoon, or a roller.

2. Brush on gel stain.

3. Pull the tape off at a 45° angle while the stain is still wet.

but be careful not to cut into the wood, or you'll get a dark line around the stencil. Peel away the tape leaving the shape of the leaf in place and stain the floor. As soon as possible after you've wiped the stain, peel away

Self-Adhering Stencils

Amazing designs are available in the form of stencils that stick to the floor. The floor shown here uses designs from Modello Designs℠ (www.modellodesigns.com); custom shapes are available.

1. Burnish both sides of the stencil before peeling the release membrane. (If you skip this step, the stencil won't release properly.)

2. Remove the waxed backing paper.

3. Align the stencil and burnish it to the floor.

4. Peel off the transfer paper.

5. Apply gel stain directly to bare wood, or paint to sealed and coated wood.

6. Remove the stencil while the stain is still wet.

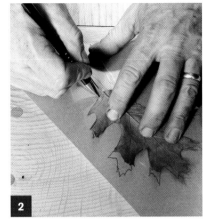

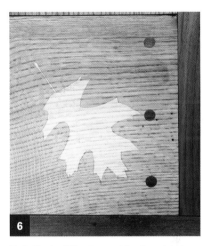

1. Burnish painter's tape to the area you want to stencil.

2. Trace the object's shape on the tape.

3. Cut out the stencil with an X-Acto knife.

4. Stain the floor.

5, 6. Peel off the tape while the stain is still wet.

the remaining tape to reveal the unstained shape. A less painstaking approach is to use self-adhering stencils, as explained in the sidebar on the facing page. Easier still are rub-on designs. All of these techniques require at least one clear finish coat atop them for protection.

DISTRESSING

Some of the homes I've worked in have literally hundreds of years of living on their floors. When a board needs replacing, or the owners want to make a new addition look old, the timeworn patina and character of the

Flooring in old houses often has centuries of character that's not easy to replicate.

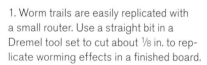

1. Worm trails are easily replicated with a small router. Use a straight bit in a Dremel tool set to cut about ⅛ in. to replicate worming effects in a finished board.

2. Stain the grooves to accent the worming area.

3. Powder-post beetle damage is common in old homes, and relatively easy to fake. Use an awl to re-create the holes.

4. Powder-post beetle holes and tracks are filled with powder. Replicate this with wood fillers. Different color fillers make the holes look as if they happened at different times.

5. Stain the board again to complete the look.

original floor are hard to match. If you want to try your hand at distressing, the effects you create should be subtle. Many craftspeople overdo their distressing. Look carefully at the existing floors, and try to spot nuances such as wear patterns and insect damage.

Remember that the old floors people find charming today were previously little more than utilitarian. Materials were expensive and hard to come by then, and grades of lumber we'd toss in the scrap heap were often used. For example, hunting was once an important way to put meat on the table, and I've occasionally found old bullets from a missed shot that hit a tree peeking from an ancient floor (see the sidebar below).

Setting a Faux Musket Ball

Embedding a piece of buckshot or a ball meant for a muzzle-loading rifle in a floor is a distinctly old touch. The soft lead of the bullet wears with the floor.

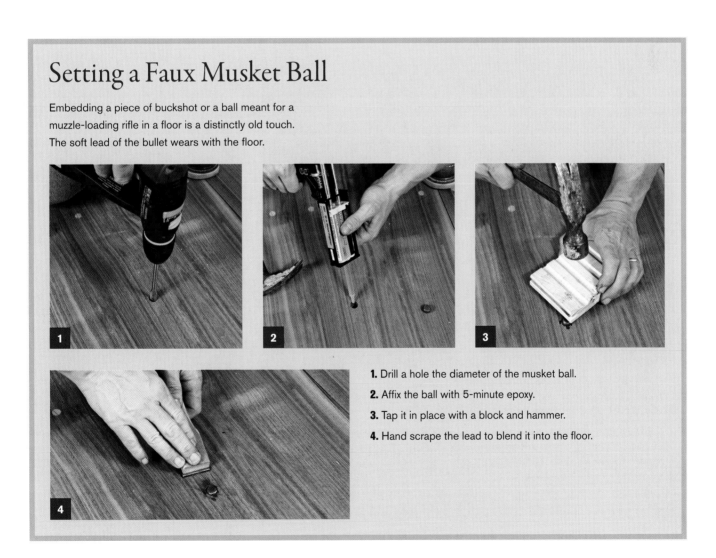

1. Drill a hole the diameter of the musket ball.

2. Affix the ball with 5-minute epoxy.

3. Tap it in place with a block and hammer.

4. Hand scrape the lead to blend it into the floor.

Hand Scraping

Back in the day, neither sandpaper nor sanding machines existed. Floor finishing was done by hand using scrapers. Most floors that are hand scraped today are done to create an aged, distressed look. Making a floor look gently worn over time, almost like driftwood found on the beach, takes patience and practice.

The soft springwood on the flooring boards wears the most and the hard end-grain knots the least. The traffic areas of the home should have more wear. On stairs, notice where your feet fall on the treads and hollow those spots out slightly.

Scrape around knots, leaving the harder, longer wearing end grain gently proud of the surrounding floor.

Scrape with the grain of the wood.

Fill knotholes with 5-minute epoxy. Both 3M's black 420 epoxy and their clear epoxy work well. Clear epoxy allows the color of the knot itself to show through.

These things become very apparent after you spend a week or two on your hands and knees scraping the floor of a 200-year-old home.

Wood species such as oak, pine, and walnut distress easily. Maple and other hard species of wood are more difficult. Keep in mind that old lumber and new lumber aren't going to look alike, even if they're the same species. Old lumber mainly came from old-growth trees that grew slowly in shaded forests. The growth rings will be tight, and the heart-wood loaded with colorful extractives. New lumber usually has more favorable growing conditions, and the growth rings are broader, the heartwood not so character laden. There are companies that sell flooring made from reclaimed lumber, and it may be worth looking into them.

When there's a lot of distressing of new lumber to be done, many flooring professionals perform most of it back at their shop on a bench. This is more efficient, but some hand scraping will need to done after installing the floor in order to blend the boards together. Flooring professionals many times will use electric grinders with wire wheels to remove soft springwood quickly and then follow by hand scraping the area. Soldering irons may be used to put in burn marks. Dark stains are applied to accent holes. Ice picks or awls are used to create insect holes. No matter what tools are used, distressing a wood floor is labor-intensive and you may only be able to accomplish 50 sq. ft. a day.

Maintaining Floors

This part is simple. Keep wood floors clean. Vacuum grit from them regularly (finish manufacturers recommend daily) and use the cleaner recommended by the manufacturer. Many manufacturers sell propri-etary finishes, and why wouldn't you use them, knowing they're formu-lated to cause no harm?

By now, you know not to wax floors that you don't want to keep on waxing forever. Keep your pet's toenails trimmed short, put a dust mat by the entries, and maybe even leave your shoes and their embedded grit on the mats. And if you allow high heels on a soft pine floor, expect there to be dents—or, should I say, character marks.

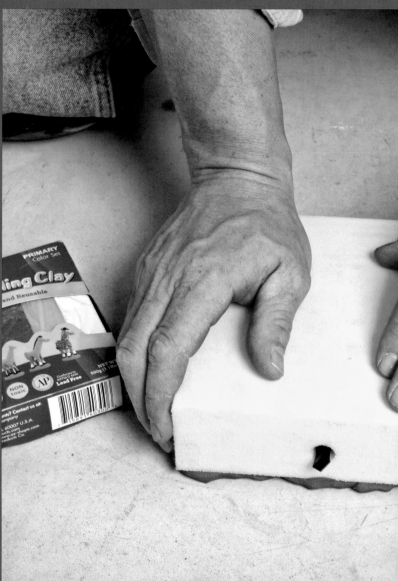

CHAPTER TWELVE

Flooring Problems & Solutions

EVERY WOOD FLOOR I INSTALL IS A LABOR OF love that requires a substantial investment in material and labor. There is no reason why it should not last the life of the home. If you install and finish floors as detailed in the previous chapters, you should never have a problem either. But not everyone is that careful, and the annual losses associated with flooring problems are approximately one billion dollars.

When wood floors have problems, they can cost a lot of money to fix—enough to bankrupt a small contractor. In this final chapter, I'll help you understand what can go wrong. Maybe you bought an existing house whose flooring is in rough shape, or

TOP LEFT Wood flooring swells and shrinks with normal changes in humidity. Install the flooring at too low a moisture content and it's likely to shrink and may develop cracks.

TOP RIGHT Cupping occurs because of a moisture imbalance across wood flooring where the bottom is wetter than the top.

BOTTOM LEFT Wood has a different dimension for every level of moisture content. Installed too dry, and without clearance to obstructions, wood that swells as it gains moisture can buckle from the subfloor.

you're a pro looking for advice about how to handle a particular job. Or maybe you're just curious about what happens if you take shortcuts on a floor installation or finishing job. In all these cases, this chapter has something for you.

Dealing with Changes in Moisture Content

As explained in chapter 1, wood shrinks and swells with changes in moisture content. Proper attention to moisture—using moisture meters to assess the job and the material, correctly acclimating wood flooring, and avoiding installations in cases where the building has chronic moisture issues—would prevent 90% of all wood flooring problems.

Wood flooring is continuously absorbing or liberating moisture, trying to equalize with its environment. Wood expands when it absorbs moisture and shrinks when it loses moisture. Install wood flooring with too high a moisture content, and you'll end up with big gaps between the boards when the flooring dries out. Installing wood flooring that's too dry for its location can have worse consequences. When overly dry flooring absorbs moisture from its environment, it could start to cup or even lift up from the subflooring.

MOISTURE METERS

I never install a floor without knowing the moisture levels of the underlying framing or concrete and the flooring itself. To determine this, I use moisture meters and relative humidity meters.

As explained in chapter 2, the two main types of moisture meters are resistance meters and dielectric meters. Resistance meters, which are about twice as accurate as dielectric meters, record moisture content by measuring how well wood conducts or resists small electrical currents. Electrical resistance is measured in ohms. The higher the number of ohms, the less conductive the material. Moisture meters are essentially ohm-meters whose readings are displayed as moisture content rather than ohms.

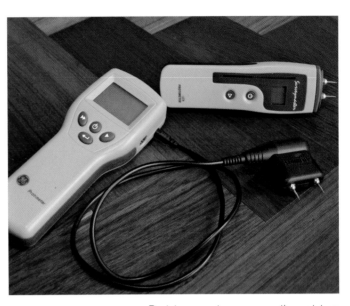

Resistance meters measure the moisture content of wood by seeing how well it conducts a small electrical current between two electrodes. Dielectric meters use surface electrodes to generate a radio field that penetrates the wood.

Wood is a poor conductor. Water is a good conductor (at least, water that contains dissolved minerals such as those found in wood; pure water is actually a poor conductor). As the moisture content of the wood increases, its conductivity increases and its resistance decreases. By measuring how well wood conducts electricity, it's possible to determine its moisture content.

Wood's electrical conductivity decreases drastically below the fiber saturation point, when the space in the center of the cells has dried out. This is also the point at which wood begins to shrink because any further drying comes from within the cell walls, which shrinks the cells. It's also the range of moisture that most concerns flooring installers. The fiber saturation point is generally between 25 and 30% moisture content. Douglas fir, for example, has a resistance of 22,400 mega-ohms at 7% moisture content and 0.46 mega-ohms at 25% moisture content. (One mega-ohm equals one million ohms.)

RESISTANCE METERS Because they have different resistances to electrical current, different wood species will indicate different moisture levels on a moisture meter even when they are at the same moisture con-

TEMPERATURE AFFECTS METER READINGS

The temperature of the wood (not the air temperature) significantly influences the readings on a resistance meter. Most meters are calibrated at 70°F. To correct for different temperatures, find the temperature of the wood in the left margin and follow horizontally to the cell corresponding to the meter reading at the top of the chart.

TEMPERATURE		METER READING					
°C	°F	6	7	10	15	20	25
-	-	ACTUAL MOISTURE CONTENT					
5	40	7	8	12	18	24	30
15	60	6	7	11	16	21	27
30	80	6	7	9	14	19	23

Temperature corrections for use with resistance-type moisture meters (adapted from *Electric Moisture Meters for Wood*, W.L. James, 1998)

tent. For example, a piece of red oak might read 8%, whereas ash at the same moisture content reads 10% and teak reads 7.5%. Moisture meter manufacturers provide a conversion chart. Most resistance meters use pine, Douglas fir, or hemlock for a calibration point. To ensure accuracy, calibrate meters regularly to a known sample.

The range of resistance meters is limited by the extremely low conductance of very dry wood. The upper limit of accuracy corresponds to the fiber saturation point of wood. The range of moisture content that can be measured with reliability by resistance-type meters is from 7 to 27%. Temperature also affects how resistance meters read moisture.

When using a resistance meter, most manufacturers recommend placing the pins parallel with the wood grain. This isn't as important for meter readings under 15%, but with moisture contents above 20%, the difference can be as high as 2%.

Any chemicals or contaminants can affect the reliability of meter readings. Extractives in the wood, such as water-soluble electrolytes, may affect readings. Although you can't always tell by looking which parts of the wood are high in extractives, it's a good bet that areas that are different in color from most of the board have greater or lesser amounts of extractives. Place the pins in areas whose color is representative of the whole board.

Using Meters with Insulated Pins

Most resistance meters come with uninsulated pins. No matter where they are placed, they will read the wettest layer of the wood. This works fine most of the time because moisture in acclimated wood is distributed fairly evenly. Sometimes, however, it's important to know the gradient of the moisture content. For example, say you're planning to sand a previously flooded floor that cups slightly. It's possible that the top of the flooring dried out more than the bottom, but there's no way to know this with bare pins. If you sanded that floor and the moisture content continued to equalize, the sanded boards could eventually crown upward. Insulated pins are covered by a tough insulating resin except at the tip. They can be driven into the flooring (or from underneath the floor) to check moisture levels at varying depths, which provides the information necessary to determine if the moisture content of the wood is even.

Insulated pin probes driven by an integral slide hammer are available to purchase for some meters. Even with the slide hammer, they may not drive easily. (Resist the temptation to drill pilot holes for the probes; if you do, the heat from the drill bit will dry out the wood in the immediate area and you won't get an accurate reading.)

If you don't have the hammer probe attachment, you can drive two finish nails into the flooring to measure the moisture content near the bottom surface. When the meter is touched to the heads of the nails, it will measure the highest moisture content reading throughout the thickness of the flooring.

Only the tips of these pins are uninsulated, allowing the probe to be driven though different layers of the floor system to provide a reading of any moisture gradient. For example, the readings might show that the top of the floor is dry but the bottom is wet.

An optional slide hammer and probe is useful for driving insulated pins deep into flooring.

Meter readings may start to drift lower after the pins are driven into wood with a high moisture content. Meter drift is less of a problem at lower moisture levels. The best practice is to take readings within 2 or 3 seconds of driving the pins into the wood. Also, the glue in plywood or engineered products can isolate layers of differing moisture content. If you think that's an issue with the wood you're checking, insulated pins can verify the accuracy of the readings. Check the sample's individual layers by driving the pins to different depths.

In dry climates, static charges may cause erratic meter readings. Minimize the effects of static charges by inserting the pins into the wood sample prior to powering up the meter. Place your hand across the inserted pins to help discharge the static charges.

DIELECTRIC MOISTURE METERS Dielectric meters used with wood usually are capacitance moisture meters, which operate on the relationship between moisture content and the dielectric constant of the wood cell. All capacitance moisture meters are based on the same operating principle. They may differ in the frequency they employ or in their electrode design. The meter sends radio frequency waves into the wood, which induces a secondary field in the wood. The meter converts the signal from this secondary field to a moisture reading. However, the moisture content of the wood isn't the only factor. Density also affects the secondary field, which is why dielectric moisture meters are less accurate than resistance meters.

On the plus side, dielectric meters do not put holes in the flooring and work quickly. They are useful for developing a general sense of moisture levels, and for locating particularly wet or dry areas where further investigation with a resistance meter is called for.

Dielectric meters can measure down to 0% moisture content with diminished accuracy. Readings above 30% moisture content are also subject to diminished accuracy. About 7 to 30% moisture content is the approximate useful range of these moisture meters. Unlike with resistance meters, temperature is not a factor for dielectric meters throughout normal ranges. Grain orientation does not matter with dielectric meters that have circular electrodes. Meters that are capable of dual-depth readings may have parallel electrodes, which should read with the grain.

Most dielectric meters use Douglas fir as a calibration point, and as with resistance meters, manufacturers supply conversion charts for other species. For me, calibration is less important with dielectric meters, as I don't use them when accuracy really counts. I just use them to find the relative moisture contents of several samples.

SUBFLOORING EQUILIBRIUM MOISTURE CONTENT

This chart shows the appropriate moisture content for subflooring at given relative humidity levels at 70°F. Don't install wood flooring until the subfloor has acclimated. The NWFA recommends a maximum difference between the subfloor moisture content and finished wood moisture content of 4% for strip flooring and 2% for plank flooring.

RELATIVE HUMIDITY	MOISTURE CONTENT (%)		
	SOLID WOOD*	PLYWOOD	OSB
10	2.5	1.2	0.8
20	4.5	2.8	1.0
30	6.2	4.6	2.0
40	7.7	5.8	3.6
50	9.2	7.0	5.2
60	11.0	8.4	6.3
70	13.1	11.1	8.9
80	16.0	15.3	13.1
90	20.5	19.4	17.2

(*Adapted from *U.S. Forest Products Laboratory's Wood Handbook*.)

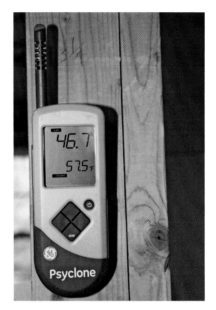

Relative humidity meters available in department stores or home centers are generally adequate; they are accurate to +/- 5% and take the reading in about half an hour. Professional meters (as shown here) are accurate to +/- 2%, and provide a reading in minutes.

MOISTURE TESTING FOR FLOORING OVER WOOD SUBFLOORING

The interior relatively humidity (RH) of most homes varies throughout the year. As the RH changes, so does the moisture content of the flooring. Wood flooring performs best if installed at a moisture content midway between the lowest and the highest moisture content expected over the course of a year.

If flooring is installed at too high a moisture content, it will dry to a moisture content approximating that of the building, resulting in shrinkage and gaps between the flooring boards. If wood flooring is installed much dryer than the average moisture content it will reach during use, swelling may occur, which can result in the complete failure of the flooring system and damage to the building structure.

Since so much depends on moisture readings, I ensure their accuracy by using multiple methods. I survey the building taking moisture content readings, RH readings, and checking the flooring's dimension. This

Where access from below is possible, check the moisture level of the joists and framing.

not only verifies accuracy but also alerts me to potential future problems.

Prior to installation, the subfloor and flooring must be at the proper moisture content expected for equilibrium with the environment. As explained in chapter 1, the proper moisture content will vary geographically. This is the wood's Equilibrium Moisture Content (EMC), which occurs when the moisture content of the wood is in balance with the relative humidity and temperature. At the EMC, wood no longer gains or loses moisture to the atmosphere. A calculator is available at www.woodflooringedu.org to convert the relative humidity and temperature readings into EMC readings. EMC readings should be calculated and compared to moisture meter readings on every project.

I like to start by taking relative humidity readings. Using a handheld electronic hygrometer, check the relative humidity and temperature of the room where the floor will be installed. Also, take readings in areas that could later have adverse effects (such as basements and crawlspaces). Electronic hygrometers take time to acclimate. You'll know that's occurred when the reading changes by no more than 0.2% over five minutes. The temperature should be within the normal limits of 60 to 80°F. Relative humidity varies by region and season, and it's recommended that installation should not take place if the interior RH is outside the normal ranges. Use the chart on p. 281 to compare the RH with the EMC of the subflooring, and the chart on p. 16. to compare it with the EMC of the flooring.

Next, I record subfloor moisture content readings. I like to use a dielectric meter to locate any areas of higher moisture content on the subfloor. You would never know without checking, for example, if someone spilled a bucket of water on the subfloor. I use a pin meter to verify the moisture content of the subfloor by taking at least three measurements per 100 sq. ft. The NWFA recommends that the maximum pre-installation difference between the subfloor moisture content and

Recording Moisture Readings

Whether you're a contractor or a homeowner, it's good insurance to record your pre-installation moisture readings. Without a record, the readings might as well not exist should you ever end up in court. A smart way to keep these records is to write them on the subfloor and the framing members where you took the reading. Use an indelible marker and take digital photos that become part of your permanent record of the job.

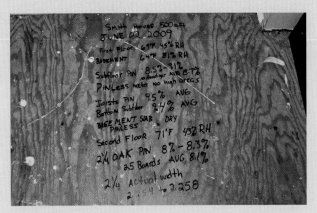

Writing the results of your moisture testing on the subfloor and taking a photo for your records provide excellent backup in the event of a warranty claim or lawsuit.

the wood flooring moisture content be no more than 4% for strip flooring and 2% for plank flooring.

If access to the floor is available from below, it's a good idea to check the bottom side of the subfloor and the joists. (If access is not available from below, optional hammer probes that get deeper into the subflooring and floor joists are available from most meter manufacturers.) Joists often harbor moisture, especially in new construction. Record your results; comparison with EMC charts will help validate subfloor moisture meter readings.

Because conditions vary throughout the year, so will the EMC. Acclimate wood flooring to the approximate midpoint of the expected seasonal moisture range. For example, in a home whose wood moisture content ranges between 6 and 12% throughout the year, wood flooring should be acclimated to about 9% at installation. Narrower boards are more forgiving in this regard. Wider flooring boards expand and shrink more than narrow ones. Because shrinking only creates gaps, whereas expansion creates cupping or even buckling, I tend to acclimate wide flooring ½% wetter than narrow flooring. This makes the wide boards more likely to shrink than to swell.

Measure the moisture content of the wood flooring to be installed using a resistance meter. Take multiple measurements and average the results.

Flooring is manufactured to specific width tolerances at given moisture contents. Verifying this information from the manufacturer prior to installation provides confirmation that the flooring was correctly milled.

The NWFA recommends measuring the moisture content of 40 boards for every 1,000 sq. ft. Do not just measure the ends of the boards that are on the top of the pile. Measure boards from the center of the flooring bundles, and from bundles in the center of the pile of flooring, if it hasn't been spread out.

I check the width of the flooring as well as the moisture content, and compare it to the manufacturer's specifications. This helps substantiate moisture content readings. When using a pin meter, push the pins in as close to their full length as possible to ensure good, consistent contact with the wood fibers.

If any of these moisture readings are outside of the correct parameters, I delay installation until the flooring and the house are at proper levels.

MOISTURE TESTING FOR FLOORING OVER CONCRETE

A lot of wood flooring gets installed over concrete slabs through direct glue down, over a floating plywood subfloor, or on some kind of sleeper system (see chapter 3). Assuming that the concrete isn't saturated with water, any of these systems can work quite well. But if the concrete has an underlying moisture problem that's not addressed, the wood floor is doomed from the start. And here's the rub: You can't tell by looking if the slab is a candidate for a wood floor. And several old-school tests don't provide reliable data. The only way to know for sure is by testing with a relative humidity meter.

Concrete always contains water, called water of hydration, which is chemically bound to the cement. Moisture readings in concrete will be quite high when compared to wood, but this is normal. Most engineered floor manufacturers require that concrete's moisture content be below 75%, or a vapor barrier is required. The trouble is that even the best test shows only the moisture content of the concrete *at the time of the test*. Who can tell if that's the wettest that concrete ever gets? I think prudence dictates the use of a vapor retarder below wood flooring on any grade level (or below) concrete slab.

That raises its own question: If I'm putting down a vapor barrier anyway, why would I bother to take moisture readings? It's a belt and suspenders thing: If I really don't want my pants to fall down in public,

I wear both belt and suspenders. I really don't want to have a job go bad, lose out on a warranty claim, or lose in court, so I test for moisture and use vapor retarders.

QUESTIONABLE CONCRETE MOISTURE TESTS What's worse than not knowing the moisture content of a slab? Thinking you do and being wrong. A commonly recommended test for concrete moisture is to tape a piece of plastic to the slab. After a minimum of 16 hours, you are supposed to inspect for moisture, discoloration, or condensation on the concrete or the underside of the plastic. The test even has a name—ASTM D4263—but it has been responsible for putting more than one flooring company out of business.

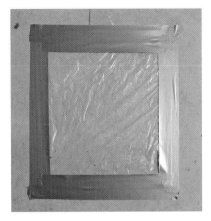

A common test for concrete slab moisture is to tape plastic to the slab and watch for condensation to develop. This test isn't of much value, however, because a lack of condensation doesn't guarantee the slab is at an acceptable moisture level.

The trouble here is that even if you don't get moisture, discoloration, or condensation, it doesn't mean the concrete is dry. It could just mean the surface temperature of the concrete is above the dew point, so no condensation occurred. You think it's OK, but the moisture level could still be too high to install wood flooring.

Another time-honored test I don't trust is the calcium chloride test. The test kit consists of a plastic dish containing 16 grams of anhydrous calcium chloride, a hydrophilic compound that absorbs moisture from the air. It uses a plastic dome of a certain volume to create a chamber to house the calcium chloride. The dome has a sealant strip that seals it to the concrete slab.

The calcium chloride, dish, lid, label, and sealing tape are weighed, and the dome is sealed to the floor over them. After 60 to 72 hours have passed, the calcium chloride dish assembly is weighed and the net weight gain calculated. From this, the concrete's vapor emission rate is calculated. Usually, this second weighing is done by a lab where the sealed test kit is sent.

There are several problems with this test. It is more sensitive to fluctuations in ambient air humidity and temperature above the slab than relative humidity testing. It measures the evaporation rate at the surface of the concrete and not the amount of free water in the concrete. The evaporation rate will vary with temperature and humidity conditions within the building. In addition, buying the test kits and sending them out to a lab for evaluation is expensive.

Dielectric moisture-meter readings are helpful in finding wetter areas to aid in determining the placement of more accurate moisture testing equipment. Do not use them to determine the acceptability of a concrete slab to receive wood flooring.

Calcium chloride testing requires cleaning a 20-in. by 20-in. square at each test site to bare concrete, followed by a 24-hour waiting period before the test kit can be placed, and it takes days to complete. When done, calcium chloride testing has measured moisture vapor emissions only from the top $\frac{1}{2}$ in. of the concrete. Finally, it will not indicate the potential concrete moisture equilibrium after installing the floor covering.

A third test is to use a dielectric moisture meter. These meters definitely have their place in concrete moisture testing, which is to give a quick read of the conditions to suggest where to do more accurate testing. I never rely on them to provide the most accurate readings.

PROBE TESTING CONCRETE FOR MOISTURE Relative humidity testing provides the best way to determine the equilibrium relative humidity in concrete. A relative humidity probe is placed in a hole drilled in the concrete. Even this test (ASTM F2170-02) tells you only the condition of the slab at that moment. It can't predict what outside influences might occur to change the moisture in the slab. What it will tell is the potential concrete moisture equilibrium after installing the floor. Always be sure the concrete has cured a minimum of 30 days before testing.

Because most new concrete slabs have a vapor retarder installed below them, they dry only from the top. To test the relative humidity of such a slab, the probe hole must penetrate 40% of the slab thickness. A 4-in. slab would require a hole depth of $1\frac{1}{2}$ in. Prior to testing, the slab must be at normal ambient living conditions for at least 48 hours. Three holes are required for the first 1,000 sq. ft., and you should do at least one more test for each additional 1,000 sq. ft. Make sure to place test holes within 3 ft. of all exterior walls for grade level or below slabs. Install a hole liner supplied by the meter manufacturer, and allow 72 hours for the holes to reach equilibrium before testing. The hole liner must be sealed in place during this time.

Once the hole has had time to reach equilibrium, insert the probe and allow it to reach ambient temperature before taking a reading. Its reading must stay within 1% relative humidity for five minutes to be accurate. Record the relative humidity, the temperature, and the location

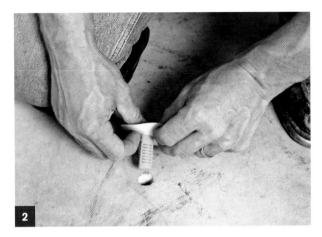

1. Holes for a relative humidity probe must penetrate 40% of the slab depth, and their diameter must not exceed 0.04 in. larger than the external diameter of the hole liner. Vacuum the dust from the hole after drilling.

2. Hole liners can be purchased from test equipment suppliers and must be composed of plastic or noncorroding metal tubes. The liner must seal to the concrete to keep out air.

3. Leave the humidity probe in the hole until the moisture level stabilizes for five minutes. The probe must incorporate a temperature sensor.

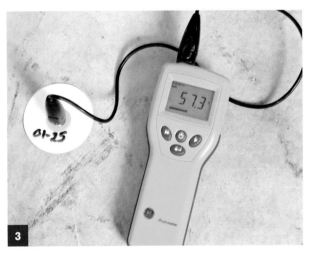

and depth of the probe. Also, make a record of the relative humidity and air temperature above the slab. Keep these records: They are your defense should there ever be a problem.

HOOD TESTING CONCRETE FOR MOISTURE Although I think the probe method is best, drilling a bunch of holes in the slab is a drawback. The hood test is ASTM recognized (ASTM F2420), and it provides an accurate measure of the free water in the concrete by testing the relative humidity in the air right above the slab. An insulated hood, available from meter manufacturers, is sealed to the top surface of a slab using, of all things, modeling clay. This creates an impermeable chamber into which a humidity probe is inserted. The probe measures the relative humidity, temperature, and dew point within the chamber. These read-

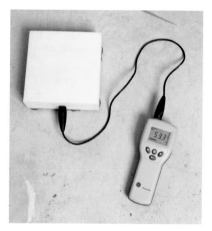

ABOVE LEFT A hood test must be performed over clean, bare concrete. Sand a small area for each test site.

ABOVE RIGHT Seal the insulated hood, available from testing equipment suppliers, to the concrete with modeling clay. Insert a relative humidity and temperature probe to provide an accurate reading of the slab's moisture level.

ings tend to be 5% lower than I get with the sleeve method, so I simply add 5% to come up with a number that I trust.

Nothing must be on the concrete slab that would interfere with moisture transmission or with the seal on the hoods, so I sand the test locations before placing them. As with hole testing, the slab must be at normal ambient living conditions for at least 48 hours before testing. The same number of tests is required—three for the first 1,000 sq. ft., and at least one for each additional 1,000 sq. ft. Tests must be conducted within 3 ft. of each exterior wall.

Give the hood 72 hours to acclimate before taking measurements. As with the hole testing, the probe must be at ambient temperature, and after the probe is inserted in the hood, the meter must remain steady within 1% relative humidity for five minutes before you record the reading. Also, record the ambient temperature and relative humidity.

Inspecting a Floor

Do not expect site-finished wood floors to resemble fine furniture. Some irregularities may be present, but they should not be prominent. For example, some sanding scratches are acceptable in particularly hard woods such as Brazilian cherry, hickory, or maple. The same is true of soft woods such as pine or fir.

What's acceptable in a refinished floor depends on the condition of the floor beforehand. Some conditions may be difficult or impossible to bring to a like-new state. The owner will have to balance the existence of defects such as stains against the cost of replacing the boards or the floor. Contractors need to be upfront with clients about such concerns so that no one is surprised.

When inspecting an existing wood floor for imperfections, do it from a standing position under normal ambient lighting. According to the standards of the National Wood Flooring Association, imperfections visible only in the glare from large windows do not count. The glare magnifies any irregularity in the floors. Both extremely light and extremely dark finishes may also accentuate irregularities. If these irregularities wouldn't be noticeable in a medium brown floor, they are not considered defects.

Problems and Solutions

In this section are the most common complaints I come across when inspecting wood floors, along with solutions for each of them. I have categorized these (alphabetically) as finish problems, sanding problems, stains, and structural problems. Gaps and cupping are by far the most common problems I see; both these conditions can be attributed to moisture problems.

FINISH PROBLEMS

When used as intended, modern wood floor finishes almost always do a good job. Be sure to read and follow the manufacturer's instructions not only during application but, equally important, also during sanding and preparation. Many finish problems relate to working in the wrong conditions—too warm, too cold, or too humid. It's far better to wait for the right day than it is to have to redo the finish.

ALLIGATORING

CAUSE The surface of the finish doesn't properly coalesce, which could be due to contaminants in the finish, too heavy a coat of finish, or working in conditions that are too cold. Other possible causes are recoating before the underlying finish has dried, especially when using oil-based and water-based finishes together, or adding excessive amounts of thinner.

FIX Screen the floor flat after the finish has fully cured and recoat.

An alligator finish pulls away from itself, gathering into a series of ridges as it dries, and is rough to the touch.

Applicator streaks: Cross-grain marks in the finish can be from the applicator; they most often occur in water-based finishes.

APPLICATOR STREAKS

CAUSE Finishes that dry before they level out can show the texture of whatever tool was used to apply them. Overly fast drying happens most often with waterborne finishes, and is exacerbated by too much air movement, working in excessively hot conditions, or the finish drying in direct sun. Other causes include not applying enough finish, applying it unevenly, or not keeping a wet edge. An applicator with hardened spots can leave marks. Flattening agents in satin or semi-gloss finishes can remain at the bottom of the can if not properly stirred, which can accentuate applicator marks. Stopping and starting the applicator in the middle of the floor also allows flattening agents to settle out and leave a mark.

FIX Screen and recoat after the finish has dried sufficiently. Sometimes a complete re-sanding to bare wood is necessary. The difficulty will be getting the floor flat. Any ridges left after screening will telegraph through the fresh finish. To check for flat, run a wet towel across the sanded floor and look across it toward the light, which will highlight any remaining ridges. The more rigid the sanding screen backer, the flatter the floor will become. Darker colored pads are generally stiffer than white pads. Whereas a white pad will flex over the ridges, dark pads keep the screen flatter, cutting off only the tops of the ridges and leveling the floor faster.

BLACK DOTS

CAUSE Common causes for black dots include iron filings that react with tannins and water, as on an oak floor finished with water-based urethane, and leave black stains. The filings can come from sharpening scrapers or a sander hitting a nail or a heating duct boot. Tannins alone can create dots, found at the end of a capillary tube in the wood. Mold can also leave black dots.

FIX Avoid problems with iron filings by not sharpening scrapers over the floor and being certain that all ferrous metal is recessed so it won't be sanded. Scrape and spot-finish black dots, recoating if large areas are involved. If the black dots are mold, odds are there's a moisture problem. Verify moisture content with a moisture meter. When the floor is dried,

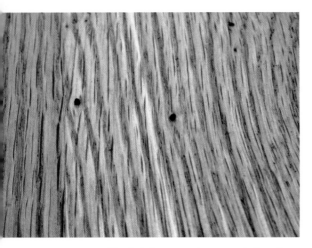

Black dots are most commonly found on tannin-rich woods such as oak, but they can occur in other species depending on the cause.

sand to bare wood, taking appropriate precautions to avoid excessive exposure to mold spores, and refinish.

BLEED-BACK

CAUSE There are many possible causes: Excessive stain soaks into the cracks between boards; highly pigmented or very thick stain doesn't dry before coating with finish; overly dry conditions at the time of staining dry the surface of the stain, trapping liquid stain below; solvents from the finish dissolve the excess stain and bring it to the surface.

FIX If the bleed-back occurs before applying the sealer coat, wipe the excess stain or buff the floor with a white pad. Allow additional drying time. If bleed-back happens after applying one or more finish coats, complete re-sanding is required. Trowel filling a floor before staining keeps excess stain from between the boards. Buffing on the stain avoids excess application. Use a fast-drying sealer to minimize the chance of re-dissolving the stain.

Bleed-back occurs when stain seeps up from the grain or from the spaces between boards.

BUBBLES

CAUSE Contaminants such as soap on the surface of the floor can cause bubbles. Other causes are overworking the finish with the applicator or not allowing bubbles from mixing to escape before applying the finish. Applying a hot finish to a cold floor causes a condition where the top of the finish skins over quickly, trapping expanding air from the wood pores that creates bubbles. Excessive air movement can have a similar effect.

FIX Bubbles are hollow and can be distinguished from bumps (see photo at right) by slicing them through the middle with a razor. Remove a few bubbles with a sharp scraper and touch fresh finish into the holes. More numerous bubbles call for screening and recoating. If surface contamination was the cause, screening may not be enough and the bubbles will reoccur in the recoat. In that case, sand the floor down to fresh wood before recoating again.

Hollow bubbles of dried finish dot the surface.

BUMPS

CAUSE Drips and dirt are two causes of bumps in the finish. Another likely cause is using unstrained finish from a partially filled can. In such

Bumps in the finish differ from bubbles in that they are solid.

cases, it's common for the finish to have partially skinned over and bits of the skin can be mixed into the finish. Two-component finishes can become lumpy if not properly mixed, as can water-borne finishes that have been frozen. Sometimes oil-based finishes develop bubbles that pop but don't fully level out.

FIX A few bumps can be scraped and touched in. If there are numerous bumps, they should be screened and the floor recoated.

CLOUDY FINISH

CAUSE Most often, this is caused by applying a coat of finish before the underlying coat has completely dried. This traps the unevaporated solvents in the lower coat. It can also occur if you use finishes from different manufacturers in successive coats.

FIX Screen and recoat. Make sure to allow adequate drying time between coats. Check with a damp rag before reapplication of finish to make sure the cloudiness has disappeared—if the finish appears clear when dampened, the problem has probably been eliminated and the floor is ready to be coated over.

As the name suggests, a cloudy finish is a cloudy or milky appearance in the finish.

CRAWLING

CAUSE Finishes can separate into their components if they sit in the can for too long. Inadequate mixing can result in applying only some of the components of the finish, which fail to form a uniform film. Contaminants on the flooring, particularly paste wax that has seeped into the joints and been re-dissolved by the finish's solvents, can also cause crawling.

FIX Small areas can be scraped and touched in. Larger areas may require screening the floor or sanding to bare wood. If contaminants are suspected, seal the sanded floor with a shellac or water-based sealer so as not to dissolve the wax. There is no guarantee that even this will work.

Crawling occurs when the finish pulls, or "crawls," away from itself.

Testing Finish Adhesion

How do you know whether the finish you're thinking of re-coating is adequately adhered to the wood? If it isn't, you won't get by with just a screening and a recoat. The floor needs to be sanded to the wood first. What if you've just finished a floor, but for some reason have a nagging doubt about whether the finish is going to stay put? Perhaps you're using a new combination of stain and finish. There's a simple way to test finish adhesion. Find an inconspicuous spot and try to pull up the finish deliberately.

Select an area free of blemishes and use a razor knife to crosshatch a pattern spaced $5/64$ in. apart down to bare wood. The test area should be approximately $3/4$ in. by $3/4$ in. Gently brush off the test area.

Place packing tape over the test area and rub it down with finger pressure. Pull off the tape at approximately a 180° angle. If small flakes of finish detach on no more than 20% of the test area, the adhesion is good. If flakes detach over 25 to 35% of the area, the adhesion is fair. More than that indicates a floor that should be sanded to bare wood before refinishing.

DEBRIS IN FINISH

CAUSE Wet finish is like a large piece of fly paper: Any dust or animal hair in the home will find its way onto it. Once the finish dries, it acts like a magnifying glass to highlight any contamination.

FIX Small amounts of contamination can be removed by spot sanding and touching in new finish. Larger areas require screening and recoating. Be certain to clean properly prior to recoating, dusting the room from the top down and including light fixtures and window trim. Tack the floor before coating. Strain even new finish before use to remove any sediment. If the finish is poured into a tray, a clean, inside-out garbage bag makes a good liner. Clean the finish applicators of any loose fibers.

Debris in the finish occurs when dirt, dust, hair, or other particles stick in the drying finish.

Discoloration: A hardwood floor changes color unevenly depending on its exposure to sunlight.

Early finish wear: Scratches, dullness, and ground-in dirt on a relatively new finish.

Grain raise: The wood surface is ridged and rough.

DISCOLORATION

CAUSE It is normal for oil-based finishes to amber over time, and most woods change color with oxidation and exposure to sunlight.

FIX If ambering is objectionable, sand to bare wood and recoat with water-based finish. Coating an oil-based finish with a waterborne one will not prevent the oil-based finish from ambering. Wood changing color over time is unavoidable, but minimize the rate of change by shading windows and skylights.

EARLY FINISH WEAR

CAUSE Early finish wear is usually a maintenance issue. Regular cleaning is required to remove grit from the floor, and pet nails must be trimmed short. Final sanding at too coarse a grit can also cause early wear, as the finish wears off the tops of the ridges left by the coarse abrasive. Improperly cured finishes can take months to harden, diminishing their wear properties.

FIX Screen and recoat. The floor may require full re-sanding if scratches and stain are excessive. Institute proper maintenance procedures.

GRAIN RAISE

CAUSE Grain raise is often caused by water getting onto the wood after sanding. The technique of water popping does this deliberately (see the sidebar on p. 258), and the raised grain is sanded smooth. Grain raise can also occur if sanding is hurried or too many grits are skipped. Waterborne finish will raise grain, as can excessive environmental moisture.

FIX Screening may flatten the grain sufficiently to yield a good recoat. If not, sand to bare wood and recoat. If grain raise is a new condition on an existing floor, environmental moisture is likely the problem. Find and cure the source of the moisture problem and ascertain that the flooring and framing are within acceptable limits before refinishing the floor.

ORANGE PEEL

CAUSE Orange peel has four common causes. Contamination such as furnish polish or wax will cause orange peel. Too heavy a coat of finish can orange-peel when the top cures long before the lower finish. As the lower finish eventually cures, it shrinks and pulls on the upper film. Applying finish to a layer that hasn't completely dried results in orange peel when the solvent from the new coat reactivates the underlying finish and causes it to wrinkle. Finally, rollers used with a finish that's drying too fast can leave this kind of texture.

FIX When the finish has hardened completely, screen and recoat. Thoroughly mix the finish and apply at the proper thickness. If rolling a fast-drying water-based finish, be sure to do so in conditions that allow the finish to flow out before drying. Contaminated floors will need re-sanding to bare wood.

Orange peel: The finish surface has the texture of an orange peel.

PEELING FINISH

CAUSE The most common cause is improper staining technique. Applying more than one coat of penetrating stain can interfere with finish adhesion, as can failing to remove excess stain residue and not allowing stain to dry thoroughly prior to coating the floor with finish. Additional causes include wax contamination, sanding with too fine a grit, inadequate cleaning between finish coats, incompatible finishes, inadequate abrasion between finish coats, and coating over a finish not thoroughly cured.

FIX If the peeling is between coats, screen the floor to sound finish and recoat. If the finish is peeling from the wood, re-sand the floor to the grit recommended by the finish manufacturer and refinish.

POLY BEADS

Beads of uncured urethane finish appear between the wood joints.

CAUSE Polyurethane droplets appear along the flooring edges when finish flows into gaps between the flooring boards at a time when the floor is expanding from a seasonal change in humidity and drying conditions are poor. The expanding boards force out uncured polyurethane. Stepping on the beads will smear them.

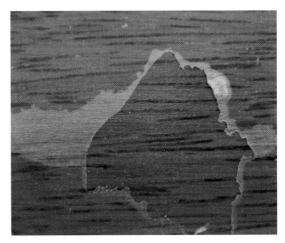

Peeling finish: The finish separates in sheets or strips from the flooring or from an underlying coat of finish.

Poly beads are a sign that the floor is moving.

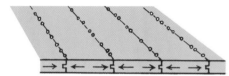

FIX If caught in time, poly beads can be removed with a razor blade, rag, and solvent. If the beads have smeared, the floor may require screening and another coat of finish. Remove all the poly beads by hand prior to starting abrading for the next coat of finish.

SCRATCHES IN THE FINISH

CAUSE Shallow scratches in the finish are a maintenance issue and are not caused by the finishing process. Neither flooring installers nor prefinished flooring manufacturers warranty against such scratches. Furniture with unpadded feet, chair wheels, grit from shoes, and pet nails can all cause such scratches. Additionally, traffic, furniture, and rugs placed on the floor before the finish fully cures can make scratches that wouldn't occur once it hardens.

FIX Screen and recoat for light scratches. Single boards may be sanded and recoated. Re-sand to the wood for deeper scratches. Although most finishes feel dry within hours, it may take a week to dry, depending on the finish and the conditions for full use. It can take weeks or even a month before placing furniture and rugs.

Scratches in prefinished floors can sometimes be touched up with kits available from the manufacturer. Otherwise, board replacement may be the best option.

SILICA STAINS

CAUSE Brazilian cherry takes up silica from the soil. Sometimes the silica bleeds out from the wood causing white marks.

Scratches in the finish are superficial scratches that don't go down to the wood.

Silica stains are white stains below the finish on Brazilian cherry.

White lines at board edges occur when the finish stretches at the edges and causes visible lines.

FIX There is no way to predict if this condition will occur. Accept it as a characteristic, or replace the board.

WHITE LINES AT BOARD EDGES

CAUSE Finish crossing the joints stretches instead of fracturing as the boards move.

FIX Screen and recoat; sanding to bare wood may be necessary. Sealer may help to prevent this.

SANDING PROBLEMS

Most sanding marks come from skipping more than one grit size per sequence. Fine sandpaper is not very efficient at removing the deep scratches left by the larger abrasive particles of rougher grits.

Improper operation of sanding equipment can lead to everything from drum marks to deep gouges from the edger. Even small details such as a slightly protruding nail head can affect the final sanding. The nail head will remove a band of abrasive around the sanding belt, which in turn will create a line in the floor with each pass.

CORNERS NOT SCRAPED

CAUSE Laziness, ignorance of hand scraping.

FIX Use hand scrapers to level the corners with the sanded floor, re-sand, and recoat.

DISHING

CAUSE Improper sanding techniques remove more material from softer areas of the floor than from harder ones. Dishing most often occurs between annual rings or between mixed species of varying hardness used together on a floor, such as in feature strips, borders, and medallions. Specifically, dishing is caused by sanding at the wrong angle to the grain, using the wrong equipment to sand mixed species floors, or screening the bare wood using too thick a driving pad.

FIX Re-sand the floor using a slight angle with the big machine. A hard plate or multi-disk sander may be necessary on softer woods.

Corners not scraped occur when areas where the machines cannot reach are not sanded.

Dishing: Softer areas of the floorboards are lower than harder ones.

Drum marks are hollows in the floor that mirror the shape of the sanding drum.

Edger marks: Circular scratches from the edger are visible in the wood.

Picture framing occurs when the perimeter of the room finishes differently than its center.

DRUM MARKS

CAUSE The drum must be raised prior to stopping the sander or a deep mark will be made.

FIX Shallower drum marks may be taken out by sanding at angles to the mark. Do not let the machine's wheels ride over the divot or new damage can be caused by the sanding drum. Once the majority of the mark is gone, finish sanding out the mark in the direction of the grain. You will have created a low spot, which must be feathered over a large area.

EDGER MARKS

CAUSE These marks are made by the edger when the sandpaper grit sequence was not adhered to.

FIX Re-sand and recoat, being sure to clean the floor thoroughly between grits. Blend the edged area into the main floor by screening or with an orbital sander before finishing. Align edger to sand with grain.

PICTURE FRAMING

CAUSE The sanding equipment used around the perimeter of the room closed off the pores of the wood, which results in the flooring taking the stain or finish differently. (This problem is also known as "halo.") Sanding the entire room with the buffer will generally blend the different techniques used on the floor together.

FIX Re-sand and refinish. Be sure to tie the edges and the center together with a final screening before applying finish.

SCREEN SANDING MARKS

CAUSE Using the old practice of sanding between coats with worn abrasive screens. In order to comply with VOC laws, new finishes are unlike the old ones so they dry differently. The abrasive screens can leave circular spider web scratches throughout the floor. Every succeeding coat of finish magnifies the scratches.

FIX Sand the finish past the coat with the initial scratches, and then continue finishing. Sand between coats with abrasive pads and sanding strips. This combination leaves smaller, more plentiful, but less noticeable scratches, which promote better adhesion between coats. Use 150-grit to 180-grit sandpaper strips for oil-based urethane and 220 grit for water-based urethane.

WAVE AND CHATTER MARKS

CAUSE Chatter marks go across the grain of the wood in a regular pattern that's usually less than 1 in. apart. There are many possible causes, including a sanding drum that's out of round or balance or that has flat spots. Bad splices on the sanding belt can also cause chatter, as can vibration from worn drive or fan belts or pulleys or bad bearings. Running the machine in the wrong direction, so its wheels aren't on freshly sanded wood, can also cause chatter.

Waves also go cross-grain but are spaced farther apart than chatter. They can be caused by out-of-round wheels on the big machine, but more likely they're just following imperfections in the floor itself, often caused by the joists or sagging subfloor. Occasionally, waves can be caused by voltage to the big machine that's too high or too low, causing it to run unevenly.

FIX Repair the big machine, if that's found to be the problem. Fix any underlying structural issues. Re-sanding the floor using the big machine at an angle to the flooring can flatten the floor. Using a buffer with a hard plate will do the same. Finally, you can use a multi-disk sander, moving slowly and lapping the preceding pass.

STAINS

Stains may be limited to the flooring finish or go deep into the wood fibers. They usually result from spills, pet urine, mildew from an underlying moisture issue, iron reacting with tannins in the presence of water (black stains found in oak), or the use of improper cleaners.

Surface stains such as residue from an improper cleaner can often be removed by using an appropriate cleaner. Some surface stains can be buffed out, but others require screening and recoating. Stains in the wood itself are more challenging. Some can be removed chemically, but

Screen sanding marks are spider web scratches in the finish.

Waves from the sander undulate across the floor.

Iron stains are visible near the nail locations.

doing so requires caution when working with the chemicals and will probably alter the color of the wood. In many cases, the best solution is to sand, treat the undesired stain, and then stain or dye the entire floor to even out the color differences before recoating.

IRON STAINS

CAUSE In the presence of water, ferrous metals react with tannins in wood such as oak, cherry, or hemlock and create black stains.

FIX Sand to bare wood and bleach with oxalic acid. Neutralize per the instructions from the oxalic acid supplier, screen, and recoat.

MILDEW STAINS

CAUSE Mildew stains point to an underlying moisture problem that must be addressed before the stain can be repaired or the wood replaced.

FIX Once the root of the stain is addressed, the first step is sanding to bare wood. This may be enough to remove the mildew stain, or it may prove adequate with finish staining of the floor. Wood bleach may also help.

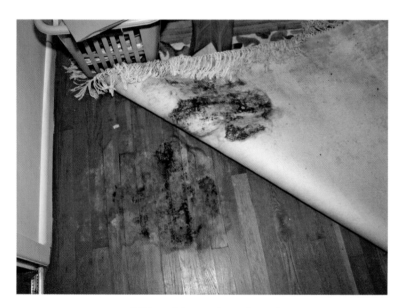

Mildew stains require prolonged levels of high moisture.

PET STAINS

CAUSE Pet urine.

FIX The best fix for stains that go below the finish, particularly from pet urine, is board replacement. Phosphoric acid (also known as naval gel) is often used with some success on small pet stains—it works by wicking tannins from the area. Common hydrogen peroxide or household chlorine bleach (don't mix the two) may also help with small stains.

Two-component wood bleach is the normal approach when an entire floor has pet stains. The first component is sodium hydroxide and the second component is hydrogen peroxide. When combined, they form sodium hydroperoxide, which is highly aggressive. It does break down wood fibers to some ex-

A pet-stained floor.

The same floor after application of two-part bleach and a medium-tone oil stain.

tent, making the floor more susceptible to denting. Neutralize the bleach according to the manufacturer's instructions, screen the floor after it dries, stain or dye it, and recoat.

STICKER STAIN

CAUSE Sticker stain is caused by wood strips (stickers) used to separate roughsawn lumber during the drying process.

FIX Sticker stain goes surprisingly deep and may not sand out entirely. It is not considered a defect in second-and-better-grade maple, number 1 common oak, and grades below.

STRUCTURAL PROBLEMS

One of the most common causes of structural problems with wood floors is moisture. What's disturbing is that moisture problems are easily avoided. If flooring installers used moisture meters correctly, acclimated flooring correctly before installing it, and refused to install flooring in buildings where chronic moisture problems went unaddressed, 90% of structural problems could be avoided.

Sticker stain: Cross-grain stripes about 1 in. wide found at regular intervals on boards.

ADHESIVE FAILURE

CAUSE Flooring adhesives typically fail from one of three causes, or a combination of them. First is moisture. If the flooring is installed too wet or too dry, or its environment becomes too wet or too dry, the wood shrinkage or expansion can overcome the adhesive bond. Second is insuf-

Adhesive failure occurs when glue-down flooring separates from its subfloor or concrete slab.

ficient adhesive transfer between the substrate and the flooring. Adhesive applied with the wrong trowel won't create a sufficient bond. Third is a contaminated substrate, such as a concrete slab treated with a sealer.

FIX Remove failed flooring and verify that any remaining flooring is adequately bonded to the substrate. There should be no squeaks, and flooring whose edges are accessible should resist slight upward pressure from a flat bar. Take readings to rule out or verify moisture as a cause. If the boards are cupped, moisture is almost certainly a factor. If moisture is found to be a factor, remedy the underlying cause before reinstalling any wood flooring.

One way to tell if a concrete slab has been sealed is to place a drop of water on it. Water on a sealed slab will bead up, while on an unsealed slab it will soak in. You can also inquire as to the slab's history, but it's possible no one will be available who can confirm the presence of a sealer. Also, samples of the adhesive can be tested for contamination with Fourier Transform Infrared Spectroscopy (FTIR), which can be done by some labs. Often, the adhesive manufacturer will send out a sample for you. The fix for a sealed slab is to sand or grind it down to uncontaminated concrete, or to reinstall the wood floor on sleepers or over a floating plywood subfloor.

BUCKLING AND BRIDGING

CAUSE Several causes exist, all related to moisture that causes the flooring to expand, push against other boards or an obstruction such as a wall, and buckle upward. The flooring could have been improperly acclimated and installed at too low a moisture content. An inadequate gap may have been left between the flooring and vertical obstructions. The room's moisture content could have increased dramatically through flooding or because of a spike in seasonal moisture from the foundation.

FIX Remove the buckled boards. Do not reinstall until you've verified that the moisture content of the flooring and the framing are within acceptable limits. You may need to refasten the flooring. Be sure to leave a gap the thickness of the floor to vertical obstructions. Sand and refinish as needed.

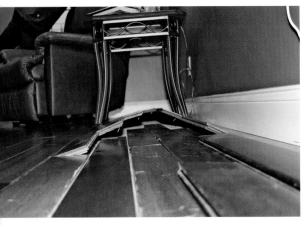

Buckling and bridging: Sections of flooring raise off the subfloor, pulling nails or breaking adhesive bonds.

CROWNING

CAUSE A moisture imbalance across the flooring, where the top is wetter than the bottom, can create crowning. That's not the most likely circumstance though. Most of the time, crowning occurs when a floor that was cupped was sanded flat. Cupping occurs when flooring is installed at too high a moisture content and its top dries faster than its bottom, or because the bottom soaked up moisture from below. When cupped boards sanded flat dry out, their center ends up crowned.

FIX When testing confirms that the floor is at its EMC, sand flat and refinish.

CUPPING

CAUSE Cupping is almost always caused by a moisture imbalance between the top and bottom of the floor boards. The bottom is wetter than the top, so it swells and cups the board. Occasionally caused by extremely

Centers of crowned floorboards are higher than the edges.

FROM CUPPED TO CROWNED

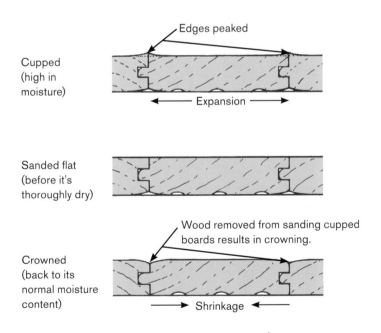

Cupped floorboards are higher at the edges than at the center.

Delamination: The veneer layer separates from the substrate of engineered flooring.

Dents are areas of crushed wood fibers.

dry conditions above, but the most likely cause is damp conditions below.

FIX Cupping of up to 0.01 in. is considered a normal condition, particularly on plank floors. Cupping of 0.02 in. or more requires attention. It is often seasonal, and the boards flatten out in dry times. For more severe cases, find and remedy the cause of any high moisture conditions. Do not sand cupped floors until the moisture content of the top and bottom of the flooring are within 1%. Allow the floor to go through one complete heating season. In many cases, this will be sufficient. If not, the wood has permanently deformed and should be sanded flat and refinished. Verify that any underlying moisture conditions have been remedied before sanding.

DELAMINATION

CAUSE Possibly a manufacturing defect. Site-related causes include an overly flexible subfloor, a subfloor that's not flat, excessive heat from ducts or too-hot radiant floor systems, and excessive moisture.

FIX Verify that the site meets all the flooring manufacturer's specifications. If it does, have the flooring tested for defects by the manufacturer. If the site is out of spec, remedy the condition and install new flooring.

DENTS

CAUSE Dents are pretty self-explanatory. One cause that often doesn't occur to people is high-heeled shoes, particularly if the heel is damaged or the floor is pine or some other soft wood. Other common causes are chairs with damaged or missing tips on their legs.

FIX Dents are difficult to repair. If the wood fibers are not broken, you can try to swell the dent to the surface using a damp cloth and an electric iron. The idea is for the moisture to swell the crushed wood, but it's tough for the steam to penetrate finished wood. This method rarely works, and it can damage the finish, but it's your only shot short of re-sanding the floor or replacing the board.

Fractures on prefinished board edges: Damaged wood or finish along the tongue edge of nailed-down prefinished flooring.

Gaps between the boards in wood floors may be normal.

FRACTURES ON PREFINISHED BOARD EDGES

CAUSE Improper use of nailer.

FIX Replace boards. Use manufacturer-supplied adapters on nailers to prevent damage in the future.

GAPS

CAUSE Most gaps are caused by a loss of moisture content.

FIX Gaps that close up during the humid season are considered a normal attribute of wood flooring. Gaps tend to be wider with plank flooring than with strip flooring. On engineered flooring, the gaps tend to happen at the butt ends.

Gaps that remain open during the humid time of the year may be filled, and the floor sanded and recoated. Repair gaps during the time of year they are smallest, usually the most humid season. Gaps repaired during the dry season may leave insufficient clearance between the boards, leading to damage when the flooring expands during the humid season.

Wood filler is not recommended for gaps wider than $3/32$ in. as it will most likely fall out. Slivers of wood can be glued in to the edge of the flooring instead. Apply glue only to one side of the sliver so as not to glue multiple boards together. Flooring boards can be pulled and replaced with wider ones.

GAPS ABOVE A FLOOR BEAM

CAUSE Floor framing settles on either side of a main beam, opening a gap in the flooring.

A gap above a floor beam runs the length of the room.

Gaps from cross-pull appear in a staircase pattern.

Other gaps are the result of poor installation techniques.

FIX Floor sheathing should overlap the beam and not join atop it. The gap may be filled during the humid season, but a better solution is to straighten the floor framing on each side of the beam. However, this is often too expensive to justify.

GAPS FROM CROSS-PULL

CAUSE The subfloor panels may be pulled in opposite directions due to settling or moisture content changes, causing gaps in the flooring.

FIX Fill gaps during the most humid time or replace flooring once the subfloor has attained EMC.

GAPS FROM INSTALLATION

CAUSE Failing to keep floor boards straight during installation; installing warped boards.

FIX Replace boards, or fill gaps during the most humid time.

GAPS FROM PANELIZATION

CAUSE Finish, typically waterborne, seeps between groups of boards and glues them together. Instead of individual boards shrinking and leaving small gaps, the group shrinks as one, leaving large gaps. Gaps that occur in increments of 4 ft. or 8 ft. can generally be attributed to subfloor panels shrinking. "Skip nailing," where a budget contractor

Forensic Gap Investigation

Gaps may be abnormal when pronounced enough to be seen from a standing position during the most humid season of the year. Abnormal gaps are generally caused by flooring installed at too high a moisture content. As the flooring dries, it shrinks. A quick check is to measure across 3 ft. or 4 ft. of flooring. The eleven 3¼-in. boards in the photo shown here should measure 35¾ in. Twenty pieces of 2¼-in. strip flooring should measure 45 in. A larger measurement suggests that the boards were oversize when installed, which might indicate they were at a higher moisture content than they should have been during installation.

A vernier caliper is used to measure a group of boards.

nails only every second or third row, is another possible cause. Verify this defect by finding nail locations with a magnet.

FIX When finishing, use a sealer. Many sealers form a relatively brittle bond with the board edge that normal seasonal movement breaks, allowing the individual boards to move. Fill during the humid season.

OVERWOOD

CAUSE Overwood (or lippage) is common in flooring. On site-finished floors, overwood is sanded flush before coating so it's not usually a problem. Prefinished flooring has beveled edges to minimize its effect. The American National Standard for Engineered Wood Flooring (ANSI/HPVA EF 2002) allows for 0.012 in. to 0.025 in. of overwood depending on the grade of wood. The National Association of Home-builders allows a surprising ¹⁄₁₆ in. (0.0625 in.) of overwood before considering it a defect. Minor overwood is caused by loose-fitting tongue-and-groove joints, often combined with uneven subfloors or debris below the flooring.

FIX Overwood should not be present on a site-finished floor. Manufacturers dictate tolerances of overwood in prefinished flooring. Ask for their standard prior to purchasing flooring. Individual boards may be replaced or the entire floor sanded and finished.

Groups of floorboards with tight joints separate from adjoining groups by relatively large gaps.

Overwood is when the edge of a board is higher than an adjacent one.

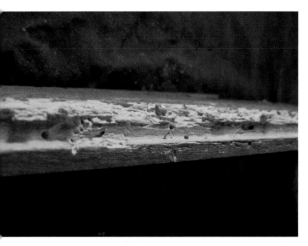
Powder-post beetles can cause severe damage to sapwood.

Punky, rotted wood can happen to the best of floors.

Shake is a defect in a board.

POWDER-POST BEETLES

CAUSE Powder-post beetle infestations are common in the sapwood of felled logs. Lumber kilns usually produce enough heat to kill the larvae, which feed on nutrient-rich sapwood until they've matured enough to emerge. It's common to find holes from old infestations in sapwood, and these are not cause for concern. Air-dried lumber may contain beetle larvae. Kiln-dried lumber may pick up beetles during storage, and beetles enter into a house in firewood, or even through open windows. Enough softwood may be destroyed to weaken the area.

FIX In an existing floor, holes from old infestations will be filled with finish. A limited number of these holes are not defects. Active infestations manifest with small "volcanoes" of wood dust appearing on the surface of the wood as beetles emerge. These can be a serious problem. Remove infested boards and consult an exterminator to minimize the risk of a house-wide infestation.

ROT

CAUSE Excessive moisture for an extended period of time, often from leaking plumbing.

FIX Address the underlying moisture problem, and allow the framing to reach moisture equilibrium. Verify with a moisture meter. Replace rotted board, and sand and coat as required.

SHAKE

CAUSE Often called wind or ring shake, this separation of the wood at the growth rings results from internal tension in the wood. It can be inherent to that piece of lumber, caused by movement in the tree from wind or from how the tree was felled. It is also thought that overly rapid kiln drying can cause shake.

FIX Cull boards with shake when installing, and be sure both the flooring and the subfloor are at the correct moisture content. Allow one full heating season for shake to appear, and replace any affected boards.

SHELLOUT

CAUSE Shellout is normally a maintenance issue. Heavy traffic and regular movement of furniture across the floor wear the softer layers of wood first. Using excessive water when mopping the floor can exacerbate normal wear.

FIX Sand and recoat. When screening, use a stiff, thin pad. A thick pad will follow the contours of the floor and may accentuate the shellout. Avoid excessive water when mopping and make sure furniture has appropriate tips or wide castors.

SPLINTERS

CAUSE When flooring is sanded one too many times, the top of the boards' groove side can become thin enough to splinter. Edges may be damaged by expansion, cupped edges may tear, subfloor irregularities may raise an edge so it can be damaged, or the grain may raise from moisture. Careless nailing may splinter an edge, or natural defects such as wind shake may predispose a board to splintering. Wire brushing may raise splinters, and bevel-edged prefinished flooring seems to be more prone than site-finished floors.

FIX If the floor has been sanded past its useful life, replace it. With luck, you may be able to replace only the worst boards. New floors sometimes produce small fibers that can be buffed away. Raised grain, cupping, and expansion damage all relate to moisture. Take care of the moisture problem and allow time for the floor to reach equilibrium. This may take a full heating season. Screen or sand as needed and recoat.

If a bevel splinters, either shave away the splinter with a razor knife and re-stain, or glue down the splinter with cyanoacrylate (Krazy Glue®). Splinters from natural defects can often be glued down in this way, too. Let the glue wick down into the seam, and then push the splinter down using something like a screwdriver that won't bond well to the glue. Don't use your finger or you can end up glued to the floor. Alternatively, replace the boards.

Shellout is wear in the softer spring or early wood.

Splinters stick out from the floor, particularly along the board edges.

Replacing a Board

Particularly with solid wood floors, board replacement isn't difficult. If you are not taking out the entire board, the first step is to make clean, square cuts at the ends of the board you're replacing. Guide a razor knife with a square to score the top of the board and prevent splintering, then make the cuts (preferably with a Fein Multimaster tool).

Use a circular saw set just to cut through the depth of the flooring to make two rips down the center of the board. Follow this up with several diagonal plunge cuts that go almost to the board's edges. These make it easier to break the board into pieces for removal. It's a good idea to locate nails first using a magnet. Use a chisel to split the board along the cuts. You should now be able to pull the board out in pieces. Be careful not to damage the tongue and groove of the adjoining boards.

Choose a replacement board whose grain is a close match to the surrounding floor. You should have the replacement board cut to size before removing the old one. Given time, floors can expand into the void when removing a board. Cut off the bottom lip of the groove of the replacement board, as well as the tongue on its end. Then bevel the bottom of the groove's upper lip to ease placing the board.

Five-minute epoxy is the best choice for gluing the replacement in place. Glue both edges to keep the board from squeaking. An alternative is a urethane construction adhesive, applied directly to the floor. The downside of using construction adhesive is that the board must be weighted down until the adhesive sets up, (say, overnight). Using a block to protect the new board, tap it down with a hammer.

1. Cut the end of the board using a Fein Multimaster tool guided with another board. If you don't have a Multimaster, it may be easier to remove the entire board.

2. Make two parallel cuts along the board with a circular saw set deep enough to reach the subfloor. Make sure there's nothing on the saw's base that could scratch the floor, and keep the cuts ½ in. away from the edges to avoid hitting nails.

3. Make several diagonal cuts along the board, being careful not to cut into adjoining boards.

4. Split the board along the cuts with a chisel.

5. Remove the board in pieces.

6. Cut off the bottom lip of the groove, and then bevel the replacement board along the bottom of the remaining lip so it can hinge into place.

7. Epoxy both edges of the board.

8. Tap in the board with a hammer and block, and weight it down as needed until the glue sets.

9. If necessary, scrape and sand the surrounding boards to bare wood to help blend the new board into the existing floor. Apply a matching stain and refinish.

SQUEAKY OR LOOSE FLOORS

CAUSE Floor squeaks are always caused by movement, either between the flooring and subfloor or in the framing. Inadequate subflooring can flex. Using too few nails in the subfloor or in the flooring can cause squeaks, as can adhesive failure.

FIX Walk across the floor and listen to locate the squeaks. Inspect the floor to determine the cause of the squeak. Loose subflooring can often be repaired from below by shimming or injecting construction adhesive between it and the joists. Loose framing can be nailed or screwed home, and joists can be stiffened by sistering on an additional member. Some floorboard squeaks can be fixed by carefully screwing through the subfloor into the flooring. Screws can be driven from above into countersunk holes, which are plugged and sanded smooth. If such noninvasive repairs aren't possible, large areas of loose flooring may need to be removed and reinstalled using correct methods.

Preserving Historic Wood Floors

Historic floors have long been a neglected subject. Floors have always been of a lower architectural hierarchy than other elements of a building, yet floors are expected to withstand centuries of wear, fire, and water damage. Conservators of historic buildings have started taking steps to preserve these fragile wood floors for posterity, but many historic wood floors have already been subjected to excessive deterioration.

Even in historic buildings that have preservation programs in place, the wood floors may not be regarded as significantly as other architectural elements. Several years ago, I toured the Biltmore Estate in Asheville, North Carolina. The Biltmore, built in 1895, is richly ornamented in the French Renaissance style and is patterned after three early-16th-century châteaux: the Blois, Chenonceau, and Chambord. An expert team of preservationists work there year-round. On my tour, I casually commented to an attendant that I believed a large area of floor had been repaired. Ten minutes later two men approached me. I tried to remember if I had walked into any restricted areas, but it turned out they just wanted to know how I could tell the floor had been repaired. To my

Properly restored historic floors complement the building, while showing marks of age.

eye, it was obvious. While they'd matched the finish well, the replacement wood was obviously new. Old wood tends to have much tighter grain, and the new boards they'd used showed wide growth rings and open grain that did not match the floors in other parts of the estate.

Clearly, replacing old floors isn't that easy, so protecting and maintaining them is a major concern. You can minimize wear by limiting traffic or placing carpet runners in walking areas. Back in the day, historic wood floors may have been cleaned with beer or vinegar or by dry washing with slightly damp fuller's earth, a fine claylike material, on the floor and dry scrubbing. Dry sweeping with herbs was also popular. The floors probably picked up some character from these cleaning techniques, but today one should use the least damaging methods to maintain the flooring. Use soft-bristle brooms or felt-lined vacuum attachments on old floors. Floors coated with varnish, shellac, or lacquer finish should never be damp mopped. If need be, use a small amount of mineral spirits to remove surface stains.

Time creates a priceless patina on the floors that is almost impossible to re-create. When possible, repair rather than replace historic wood flooring. I use historically appropriate methods when repairing old floors; for example, scraping worn finishes rather than sanding them.

Restoring historic wood flooring involves a great deal of research. Review the building's history to find why and when it achieved significance. Each historic place is a physical record of its time and the flooring should be restored to the period of significance.

TOP LEFT This floor was installed in Thomas Jefferson's parlor circa 1804. A tongue-and-groove joint holds cherry panels into beech frames on two sides, while iron pins hold the other sides. The frames join with a mitered half lap screwed together from the back. The original finish was probably oil and beeswax. Now, the floor is occasionally waxed and buffed, and any touchup is done with orange shellac.

TOP RIGHT Restoring an old floor sometimes yields treasure. This turn-of-the-century parquet had been hidden under layers of carpet and tiles.

BOTTOM LEFT This plank floor, which dates to the 1700s, is being hand scraped to restore it to the period of significance. The joints are scraped level first, followed by the middle of the planks.

Don't neglect the rest of the house when working on old floors. The walls and trim are historic, too, and need protection.

REPLACING BOARDS

When board replacement is unavoidable, duplicate the patina, wood grain, age, color, and texture as close as possible. My grandfather, who was a genius at restoration, repeated these principles to me so many times that I hear them in my sleep. If the flooring is too severely deteriorated to be repaired, try to replace with boards taken from normally hidden areas such as under stairs or cabinets. Document changes well for future reference.

Not all historic buildings contain a supply of boards from inconspicuous places. Sometimes the replacement has to come from outside. If you do much historic preservation work, it's a good idea to create a supply of aged boards. When the opportunity arises, I take them from old buildings that don't have much significance or are being torn down. Sometimes you can find old wood in architectural salvage yards. Old-growth eastern white pine was commonly used throughout the northeast and heart pine in the southern states. The eastern white pine is often referred to as pumpkin pine because it develops a warm amber patina over time. Lower grades of plank flooring were commonly used on the second floor of the home and in rooms of less importance. New lumber from tree plantations doesn't come close to replicating the look or the hardness of this old-growth lumber.

Some of the earliest softwood floors were uncoated. These bare wide planks were generally maintained by washing them with water and homemade lye. Over their life these floors may have been painted, finished with linseed oil, waxed, shellacked, or varnished (urethane is not varnish, though it's often called that), and all these finishes can be replicated today.

After grain, color can be the hardest thing to match. Some people try to age new boards with lime solutions or fuming them with ammonia, but it generally comes down to a good eye and a mixture of dyes and stains. Although modern stains and dyes are often used, a stain made from boiling walnut husks creates a lovely shade of deep golden brown.

Historic floors take a lot of work and expertise. Without the expertise, you probably won't be called on to work on significant buildings. But there's still a lot of satisfaction in restoring a true colonial or Victorian floor to its original state. What else do most of us get to do in our lives that will go on serving its original purpose for the better part of another century?

Pieces of a damaged border are removed carefully and replaced with wood of a similar age and grain pattern.

The selection of wood species available for flooring has exploded in recent years, particularly in engineered products. No longer do installers have only oak and maple to choose from.

Listed here are some of the most common flooring species, along with a brief description of their working characteristics. Since red oak is still the most commonly used species, the hardness of all other species is expressed in relation to red oak. The relative stability of each species to red oak is based on a comparison of dimensional coefficients.

Also, specific health risks are listed here, but keep in mind that regular exposure to wood dust of any species, particularly through inhalation, increases your chances of developing chronic lung conditions and certain cancers. Use dust collection and wear appropriate respirators and masks particularly when sanding.

Ash, white

ORIGIN North America

COLOR Light brown heartwood with creamy white sapwood

COLOR CHANGE Medium degree of color change; ambers to a straw tan as it ages

GRAIN Straight, moderately open grain

RELATIVE HARDNESS 2% harder than northern red oak

RELATIVE STABILITY 26% more stable than northern red oak

NAILING No known problems

SANDING No known problems

FINISHING Staining may be difficult

HEALTH RISK Respiratory irritant

Beech

ORIGIN North America

COLOR Pale white with traces of reddish brown

COLOR CHANGE Little

GRAIN Mostly closed, straight grain; fine, uniform texture

RELATIVE HARDNESS 1% harder than northern red oak

RELATIVE STABILITY 17% less stable than northern red oak

NAILING Tongues split easily

SANDING No known problems

FINISHING Staining may be difficult

HEALTH RISK Sensitizer

Birch

ORIGIN North America

COLOR There are several subspecies of birch: Yellow birch's sapwood is creamy yellow to white, heartwood is light reddish brown, with traces of red; sweet birch's sapwood is creamy white, heartwood is pale brown with shades of red

COLOR CHANGE Little

GRAIN Medium figure, straight, closed grain, even texture; some curly or wavy figure

RELATIVE HARDNESS 2% softer than northern red oak

RELATIVE STABILITY 9% more stable than northern red oak

NAILING No known problems

SANDING No known problems

FINISHING Staining may be difficult

HEALTH RISK Sensitizer

Bubinga

ORIGIN Africa

COLOR Reddish brown background with purple streaks

COLOR CHANGE Undergoes a medium color change from pinkish rose when freshly milled to burgundy red when fully aged

GRAIN Dense, fine; can be straight or interlocked, can contain figuring

RELATIVE HARDNESS 109% harder than northern red oak

RELATIVE STABILITY 2% more stable than northern red oak

NAILING Tongues split easily

SANDING No known problems

FINISHING Oil-based polyurethane can have difficulty drying

HEALTH RISK Irritant

Cherry, American

ORIGIN North America

COLOR Cream-colored sapwood with pinkish to red heartwood

COLOR CHANGE Cherry undergoes an extreme color change, darkening from pink when fresh milled to a dark reddish color when fully aged; this process occurs within a few weeks in direct sunlight and by oxidation, out of sunlight, over a six- to eight-month period

GRAIN Fine, frequently wavy, uniform texture; distinctive flake pattern on quartersawn surfaces; texture is satiny, with some gum pockets

RELATIVE HARDNESS 26% softer than northern red oak

RELATIVE STABILITY 31% more stable than northern red oak

NAILING No known problems

SANDING No known problems

FINISHING No known problems

HEALTH RISK Respiratory irritant

Cherry, Brazilian

ORIGIN South and Central America

COLOR Rich reddish to orange-brown heartwood, black streaks and grayish white sapwood; silica/white spots are a natural occurrence in Brazilian cherry and are not considered an imperfection

COLOR CHANGE Changes dramatically over time to a deep rich red; in direct sun, the color change will occur within a few days, out of sunlight, it will change over six months; water-based finishes tend to slightly mask the color change, while oil-based finishes highlight it

GRAIN Mostly interlocked; texture is medium to rather coarse

RELATIVE HARDNESS 119% harder than northern red oak

RELATIVE STABILITY 21% more stable than northern red oak, although some batches are considerably less stable

NAILING Tongues split easily

SANDING Abrasive scratches will not be removed if you skip more than one grit size per sequence

FINISHING Oil-modified polyurethane may cause white spots or specks, as well as white end joints; avoid by buffing in a clear oil sealer or neutral stain, then buffing on satin polyurethane

HEALTH RISK Irritant

Cypress, Australian

ORIGIN Australia

COLOR Honey-gold to brown heartwood with darker knots throughout; sapwood is pale cream with slight yellow tones

COLOR CHANGE Color may mute or amber modestly over time

GRAIN Closed with many knots

RELATIVE HARDNESS 7% harder than northern red oak

RELATIVE STABILITY 56% more stable than northern red oak

NAILING Tongues split easily

SANDING Tendency to clog paper due to high resin content; hardplating and screening may leave swirls; screening more than twice may be necessary; knots are extremely hard and may cause waves in the floor

FINISHING Finish adhesion issues on knots

HEALTH RISK Potential for respiratory/allergic reactions

Douglas fir

ORIGIN North America

COLOR Heartwood is orange/yellow to light brown; sapwood is tan to creamy white

COLOR CHANGE Darkens radically with sunlight

GRAIN Straight with occasional figuring

RELATIVE HARDNESS 49% softer than northern red oak

RELATIVE STABILITY 27% more stable than northern red oak

NAILING No known problems

SANDING May be resinous

FINISHING Staining may be difficult

HEALTH RISK Irritant

Hickory

ORIGIN North America

COLOR Heartwood is reddish brown; sapwood is creamy white to pinkish

COLOR CHANGE Little

GRAIN Normally straight but may have some figuring; course texture

RELATIVE HARDNESS 41% harder than northern red oak

RELATIVE STABILITY 11% less stable than northern red oak

NAILING Tongues split easily

SANDING Light color and high density make it more difficult to sanding marks

FINISHING Staining may be difficult

HEALTH RISK Irritant

Jarrah

ORIGIN Australia

COLOR Heartwood is uniformly rich reddish brown to a soft salmon pink, with black streaks and pale sapwood

COLOR CHANGE Turns a deep brownish red

GRAIN Mostly straight grain, relatively coarse but even; may have slightly interlocked grain

RELATIVE HARDNESS 48% harder than northern red oak

RELATIVE STABILITY 7% less stable than northern red oak

NAILING No known problems

SANDING No known problems

FINISHING Color may bleed into finish

HEALTH RISK Nose and throat irritant

Maple, hard

ORIGIN North America

COLOR Varies from pale cream sapwood to different tan/brown tones

COLOR CHANGE Medium color change, with slight ambering over time

GRAIN Closed, straight with occasional figuring

RELATIVE HARDNESS 12% harder than northern red oak

RELATIVE STABILITY 4% less stable than northern red oak (though in most situations changes more)

NAILING No known problems

SANDING Sanding marks and finish lines are more obvious due to maple's density and light color; maple burnishes easily, dulls fine abrasives, and may be difficult to remove previous scratches

FINISHING Staining may be difficult

HEALTH RISK Respiratory sensitizer; spalted maple dust can induce severe respiratory problems

Oak, red (northern)

ORIGIN North America

COLOR Varies from lighter tans with pinkish highlights to darker browns

COLOR CHANGE Medium change, ambering of the fresh pinkish brown over time

GRAIN Open, coarse

RELATIVE HARDNESS AND STABILITY Most common American flooring wood, baseline for all others

NAILING No known problems

SANDING No known problems

FINISHING No known problems

HEALTH RISK Eye and respiratory irritant

Oak, white

ORIGIN North America

COLOR Heartwood is light to dark brown, rarely has a pinkish hue; white to creamy sapwood

COLOR CHANGE Medium ambering over time

GRAIN Open, medium course grain

RELATIVE HARDNESS 6% softer than northern red oak

RELATIVE STABILITY 1% less stable than northern red oak

NAILING No known problems

SANDING No known problems

FINISHING Tannins may react with some water-based finishes or bleaches to turn the wood green or brown

HEALTH RISK Eye and respiratory irritant

Padauk

ORIGIN Africa, Asia

COLOR Heartwood is deep red to purple-brown with creamy sapwood

COLOR CHANGE Darkens to reddish purple-brown or black over time

GRAIN Straight to interlocked; moderately coarse texture

RELATIVE HARDNESS 34% harder than northern red oak

RELATIVE STABILITY 51% more stable than northern red oak

NAILING No known problems

SANDING No known problems

FINISHING Difficulty with oil-based polyurethane drying

HEALTH RISK Commonly a respiratory, eye, and skin irritant; may cause nausea

Pine, eastern white

ORIGIN North America

COLOR Red/brown to cream

COLOR CHANGE Slight darkening

GRAIN Closed, clear to knotty

RELATIVE HARDNESS 71% softer than northern red oak

RELATIVE STABILITY 43% more stable than northern red oak

NAILING No known problems

SANDING Easy to crush and burnish grain

FINISHING Difficult to stain

HEALTH RISK Irritant

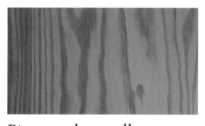

Pine, southern yellow

ORIGIN North America

COLOR Heartwood varies from orange to reddish brown, with tannish yellow sapwood

COLOR CHANGE Darkens dramatically

GRAIN Closed straight grain, course texture

RELATIVE HARDNESS 47% softer than northern red oak

RELATIVE STABILITY 27% more stable than northern red oak

NAILING No known problems

SANDING Resinous; will clog abrasives

FINISHING Difficult to stain

HEALTH RISK Eye and respiratory irritant

Purpleheart

ORIGIN South and Central America

COLOR Heartwood is brown to purple, sapwood is a lighter cream color; minerals in some boards may cause uneven color

COLOR CHANGE Extreme color change over time; freshly milled or sanded purpleheart will be brown, which changes to purple within a few days; over time, the purple oxidizes back to brown with purplish highlights; water-based finishes tend to retain more of the purple

GRAIN Straight; medium to fine texture

RELATIVE HARDNESS 124% harder than northern red oak

RELATIVE STABILITY 43% more stable than northern red oak

NAILING No known problems

SANDING No known problems

FINISHING Color may bleed with some finishes

HEALTH RISK Commonly a respiratory, eye, and skin irritant; may cause nausea

Santos mahogany

ORIGIN South America

COLOR Dark reddish/purplish brown with light orange-brown and yellow overtones

COLOR CHANGE Slight muting over time

GRAIN Fine, even-textured grain

RELATIVE HARDNESS 71% harder than northern red oak

RELATIVE STABILITY 36% more stable than northern red oak

NAILING No known problems

SANDING No known problems

FINISHING Oil-based satin finishes may have sheen variations

HEALTH RISK Respiratory irritant

Teak (Burmese)

ORIGIN Myanmar and Thailand

COLOR Pale yellows to orange-browns, with darker striping

COLOR CHANGE Extreme muting into mellow medium browns

GRAIN Straight to wavy grain; coarse texture

RELATIVE HARDNESS 22% softer than northern red oak

RELATIVE STABILITY 52% more stable than northern red oak

NAILING No known problems

SANDING Generally difficult to sand; oils clog abrasives

FINISHING Teak oils may interfere with adhesion and drying of some oil-based finishes

HEALTH RISK Common skin and respiratory irritant

Walnut, black

ORIGIN North America

COLOR Heartwood is chocolate brown, occasionally with darker, sometimes purplish streaks; sapwood is nearly tannish white; steaming the walnut will help with uniformity of color by allowing the darker heartwood pigments to bleed into the sapwood

COLOR CHANGE Exhibits a medium degree of color change; the heartwood lightens to golden brown

GRAIN Straight and open with some figuring

RELATIVE HARDNESS 22% softer than northern red oak

RELATIVE STABILITY 26% more stable than northern red oak

NAILING No known problems

SANDING No known problems

FINISHING No known problems

HEALTH RISK May cause both contact dermatitis and respiratory allergic reactions; dust toxic to horses

Wenge

ORIGIN Africa

COLOR Heartwood is dark chocolate brown, with yellowish white sapwood

COLOR CHANGE Pronounced darkening within a few months to a deep chocolate/black brown

GRAIN Straight, coarse texture

RELATIVE HARDNESS 26% harder than northern red oak

RELATIVE STABILITY 46% more stable than northern red oak

NAILING No known problems

SANDING Moderately difficult; abrasive scratches may not be removed if you skip more than one grit size per sequence

FINISHING Oil finishes may not dry; always test the finish you plan to use to ensure compatibility; staining may be difficult

HEALTH RISK Common sensitizer, respiratory, eye, and skin irritant

How to Predict Dimensional Change

Scientists have calculated dimensional change coefficients for wood in the common moisture range between 6 and 14%. These coefficients can be used to estimate how much a floor may gap or if it will buckle off the subfloor. "C_R" is the radial dimensional coefficient used with quartersawn material. "C_T" is the tangential dimensional coefficient used with plainsawn material.

For example, a 10-in.-wide plainsawn red oak plank installed at 12% moisture content will gap 0.1845 in., or almost 3/16 in., if its moisture content drops to 7%. If the moisture change was in the opposite direction, the floor might have buckled off the floor. Here's the formula for calculating dimensional change:

(percent change in moisture) × (width in inches) × (dimension coefficient) = dimensional change
(5) x (10) × (0.00369) = 0.1845 in.

HARDWOOD SPECIES	DIMENSIONAL CHANGE COEFFICIENT	
	C_R	C_T
Ash, white	0.00169	0.00274
Aspen, quaking	0.00119	0.00234
Basswood, American	0.00230	0.00330
Beech, American	0.00190	0.00431
Birch, paper	0.00219	0.00304
Birch, yellow	0.00256	0.00338
Butternut	0.00116	0.00223
Cherry, black	0.00126	0.00248
Chestnut, American	0.00116	0.00234
Elm, American	0.00144	0.00338
Hickory, pecan	0.00169	0.00315
Hickory, true	0.00259	0.00411
Holly, American	0.00165	0.00353
Madrone, Pacific	0.00194	0.00451
Maple, sugar	0.00165	0.00353
Oak, red	0.00158	0.00369
Oak, white	0.00180	0.00365
Sassafras	0.00137	0.00216
Sweet gum	0.00183	0.00365
Sycamore, American	0.00172	0.00296
Walnut, black	0.00190	0.00274
Yellow poplar	0.00158	0.00289

Dimensional Change of Selected Wood Species

SOFTWOOD SPECIES	DIMENSIONAL CHANGE COEFFICIENT	
	C_R	C_T
Bald cypress	0.00130	0.00216
Cedar, eastern red	0.00106	0.00162
Douglas-fir, coast-type	0.00165	0.00267
Pine, eastern white	0.00071	0.00212
Pine, jack	0.00126	0.00230
Pine, longleaf	0.00176	0.00263
Pine, red	0.00130	0.00252
Pine, shortleaf	0.00158	0.00271
Pine, sugar	0.00099	0.00194

(Hardwood and softwood tables adapted from Forest Products Laboratory; imported wood table adapted from Wood Flooring International.)

IMPORTED WOODS	DIMENSIONAL CHANGE COEFFICIENT	
	C_R	C_T
Afrormosia	0.00190	0.00329
Amendoim	0.00213	0.00251
Cherry, Bolivian	0.00197	0.00234
Cherry, Brazilian	0.00218	0.00289
Cherry, Patagonian	n/a	0.00302
Cypress, Australian	n/a	0.00162
Doussie	0.00138	0.00244
Iroko	0.00125	0.00192
Jarrah	0.00274	0.00236
Kempas	0.00308	0.00357
Mahogany, Royal	0.00202	0.00261
Mahogany, Santos	0.00163	0.00234
Maple, Patagonian	0.00207	0.00231
Merbau	0.00223	0.00289
Rosewood, Bolivian	0.00146	0.00205
Rosewood, Caribbean	n/a	0.00211
Rosewood, Patagonian	n/a	0.00324
Teak, Brazilian	0.00312	0.00364
Teak, true	0.00138	0.00176
Tigerwood	0.00147	0.00228
Walnut, Brazilian	0.00266	0.00307
Walnut, Caribbean	0.00291	0.00349
Walnut, Patagonian	0.00197	0.00319
Wenge	0.00235	0.00302

Affixing wood flooring with adhesive (as opposed to nails or staples) can offer some advantages. Most engineered flooring is intended to be installed this way, but adhesives can also be a great alternative for solid wood flooring.

For example, it's difficult to impossible to fasten wide-plank flooring adequately with only blind nails. Adhesive is more than capable of making up the difference.

There are a number of adhesive types on the market. Each has its uses, and one or two are nearly universally appropriate. What follows is an introduction to the main types of flooring adhesives. Because of ever-stricter VOC regulations, solvent-based adhesives may not be available in all locations and they are not discussed here.

Moisture-Cure Urethanes

Moisture-cure urethane (MCU) adhesive works well with all types of wood flooring and subfloor material. I use Bostik's Best moisture-cure urethane adhesive for all my glue-down installations. MCU adhesive cures by reacting with moisture in the air to change to a solid state. Since moisture-cure urethane contains no water, there is no chance of it causing the flooring to cup as can happen with water-based adhesives.

If initial grab strength is desired so that the flooring stays put until the adhesive cures, MCUs generally require "tack" time. Grab strength is the initial holding power of the adhesive, while tack time is the waiting period between spreading the adhesive and installing the floor. Because MCUs cure by taking small amounts of moisture from the environment, they cure faster in high humidity and more slowly in low humidity.

MCUs have a much slower grab time than their water-based counterparts, so the installer generally does not stand on the newly installed flooring while working. MCUs have a limited open time, or maximum amount of time the adhesive can be exposed to air before installing the flooring. If you exceed the open time, the adhesive will cure and be unable to bond with the flooring.

MCUs create a waterproof bond but are often represented incorrectly as moisture retarders. The adhesive may not have the required perm rating.

Most adhesive manufacturers require a specific vapor retarder system to provide acceptable moisture conditions for wood flooring.

Be aware that MCUs can etch the surface of prefinished floors that have a urethane finish. Manufacturers sell adhesive removers for their product, though mineral spirits usually work. Dry adhesive is difficult to remove. Be careful: It's possible to rub a finished floor too hard and end up burnishing the finish. This can change the sheen in the burnished area unacceptably.

Water-Based Pressure-Sensitive Adhesives

When water-based pressure-sensitive adhesives (PSAs) are first applied to the subfloor they have little grab strength. As water flashes off, the adhesive becomes tacky and remains so. It develops aggressive grab strength. Water-based adhesive is generally ready for flooring installation to start when it feels tacky but the adhesive does not transfer to your finger. Water-based adhesives generally require only soap and water for cleanup when still wet. Afterward a solvent such as mineral spirits would be required.

Slight cupping or end lifting can occur when too little time is allowed for moisture to evaporate from the adhesive prior to placing the wood flooring. Correct this by increasing the open time. However, if you exceed PSA's open time, the flooring will stick to the adhesive on contact. If you attempt to push the flooring past this point into place, the so-called adhesive memory will move the flooring back to its original placement and gaps are likely to result. Many manufacturers recommend countering this by rolling on a thin layer of adhesive. This slightly wet layer will act almost like a lubricant, allowing the flooring to slide tightly into place.

Water-based adhesives are generally used with engineered flooring and are not recommended for flooring that may be susceptible to cupping, such as bamboo or solid wood. Any water-containing adhesives applied over a nonporous substrate such as a vapor barrier will increase the chances of the flooring cupping or its ends lifting. Since no water can be absorbed into the substrate, all of it must escape upward through the flooring.

It is a common misconception that PSAs are free of solvents: Many contain solvents, just as latex paints do. These solvents are required for the emulsion to coalesce once the water evaporates. Many latex adhesives actually contain more solvent than moisture-cure urethanes.

WATER-BASED ADHESIVES AND PH

It's possible that water-based adhesives used on concrete slabs can fail due to high pH levels. Check to see if the manufacturer requires pH testing of the slab. Excessive moisture moving through a concrete slab to the top may carry soluble alkalis, increasing the pH level at the surface enough to break down or re-emulsify some adhesives.

Testing for the pH level is accomplished by placing a few drops of water on the concrete surface, waiting a given time interval, and then dipping pH test paper into the water. Immediately remove the test strip and compare to the color on the chart provided.

Alkalis require moisture, and a dry slab will not have problems. In any event, there should be a vapor retarder separating and protecting the adhesive from the concrete.

Modified Silane Polymer Adhesives

The latest generation of flooring adhesives is modified silane (MS) polymer adhesives. These adhesives were developed in Japan, are used extensively in Europe, and are now being introduced to the United States.

The polymers in MS adhesives create an elastic adhesive bond. The adhesive's high elasticity allows the flooring to return to its original position throughout the moisture cycle, which allows the floor to move with changes in humidity without developing permanent gaps. MS polymer adhesives are moisture-curing adhesives, but unlike urethane adhesives they do not contain isocyanate. They are moisture proof when dry and have quick early grab strength. MS adhesives contain no water and work well with moisture-sensitive products. However, MS generally costs more than MCU, has lower bond strengths, and slumps badly. ("Slump" refers to how much the ridges in the adhesive made by the notches in the trowel sag. Ridges that don't sag are preferred because they allow good contact to both surfaces.)

MS polymer adhesive will not etch the surface of urethane flooring finish like urethane adhesives. The adhesive can be removed from the flooring surface using mineral spirits.